At 7:17 on the
hundreds of observ
frightened by a huge, shining cylindrical object that blazed across the sky and disappeared in a flash of searing light. Moments later the earth shuddered under the impact of a cataclysmic explosion.

The initial thermal blast was followed by a hideous firestorm, and a black rain that contaminated hundreds of square miles. The afterglow turned night into day in London, and the seismic shock was felt around the world—from Moscow and Paris to Washington, D.C.

What caused it? A comet? A meteorite? A spaceship?

THE FIRE CAME BY

For nearly seventy years the question has gone unanswered. Now, in a documentary report as gripping as a detective story, author John Baxter and Thomas Atkins present the stunning explanation of the most powerful explosion mankind has ever known.

"The greatest scientific mystery of our century . . . devastation only comparable to Hiroshima." —Arthur Koestler

"A well-documented and startling investigation of an atomic explosion in the skies over Siberia—39 years before the atom bomb was 'invented.' This chilling book will bring you long thoughts and perhaps the conviction that we of Earth are not the only probers of space."

—Charles Berlitz
author of THE BERMUDA TRIANGLE

"Serious and thorough . . . enough evidence to make a plausible case that the Siberian blast was in fact an atomic explosion . . . the theory is intriguing and the question remains: if this was indeed an atomic blast, how else could it have happened?"
—*Newsweek*

"Fascinating . . . reminds us that the real universe is an infinitely more exciting and unpredictable place than both the run-of-the-mill rationalists and the hack fantasists would have us believe."
—*New York Times*

"A startling hypothesis amply established."
—*Detroit News*

"Baxter and Atkins tell the story with great skill; doing it methodically, in order and detail, taking every factor into account, and going about the process, very rationally, of making an absorbing suspense tale out of it. And in the end, they come up with a dramatic solution indeed. . . . But read the book! I loved every page of it!"
—*Isaac Asimov*

THE FIRE CAME BY

THE RIDDLE OF THE GREAT SIBERIAN EXPLOSION

John Baxter
and
Thomas Atkins

Introduction by
Isaac Asimov

A Warner Communications Company

WARNER BOOKS EDITION

Photographs 1–5, 10–14, 31–33 are from *Giant Meteorites* by E. L. Krinov, Pergamon Press, Oxford, England; 6, 8 courtesy of the American Museum of Natural History; 7, 9 courtesy of the University of Chicago Press; 16, 17 Meteor Crater Enterprises; 18 from *The Conquest of Space*, paintings by Chesley Bonestell, text by Willy Ley, copyright 1947 by Chesley Bonestell, all rights reserved, reprinted by permission, Viking Penguin Inc.; 19 courtesy of Little, Brown & Company, all rights reserved; 20 Lick Observatory; 21, 37, 39 NASA; 22, 27 Culver Pictures; 23, 24 Nevada Test Organization, Lookout Mountain Laboratory; 25, 26, 28 USAF; 29, 30 courtesy of W. W. Norton & Company; 35 *The National Enquirer;* 36 the London *Daily Express;* 38 Eros Data Center.

Excerpts from Anton Chekhov's letters are taken from *The Selected Letters of Anton Chekhov*, ed. Lillian Hellman, Farrar, Straus & Cudahy, copyright © 1955 by Lillian Hellman. For permission to reprint material from *Giant Meteorites* by E. L. Krinov, copyright © 1966 by Pergamon Press Ltd., grateful acknowledgment is made to the publishers.

Library of Congress Catalog Card Number: 75-36578

ISBN 0-446-89396-X

This Warner Books Edition is published by
arrangement with Doubleday & Company, Inc.

Cover art by Lou Feck

Warner Books, Inc., 75 Rockefeller Plaza, New York, N.Y. 10019

Printed in the United States of America

Not associated with Warner Press, Inc. of Anderson, Indiana

First Printing: May, 1977

10 9 8 7 6 5 4 3 2 1

Acknowledgments

The authors wish to express their appreciation first of all to their translator, John W. Atwell, associate professor of Russian history at Hollins College, Virginia; his expert assistance and advice on this project were invaluable. Other translations were undertaken by Laird Ekland. Next, Larry Ashmead must be thanked for proposing the idea for the book and giving us generous support and encouragement. Mary Ellen O'Brien made many useful suggestions that influenced the shape of the manuscript, and R. H. W. Dillard was helpful in matters of cosmology.

We owe a debt of gratitude to the staff of the Hollins College Library, particularly to Shirley Henn who patiently handled all of our numerous and often obscure requests for interlibrary loan articles and books. Careful typing and proofreading of the manuscript were done by Eileen Montgomery. We are also grateful to the Library of Congress in Washington, D.C., the New York City Public Library, the Library of the University of Missouri at Columbia, and the St. Louis Public Library where we conducted some of our research.

Contents

The fire came by and destroyed the forest, the reindeer, and the storehouses. Afterward, when the Tungus went in search of the herd, they found only charred reindeer carcasses.

Witness to the Siberian
explosion of 1908

I was sitting on my porch facing north when suddenly, to the northwest, there appeared a great flash of light. There was so much heat that . . . my shirt was almost burned off my back. I saw a huge fireball that covered an enormous part of the sky. . . . Afterward it became dark and at the same time I felt an explosion that threw me several feet from the porch. I lost consciousness. . . .

Witness forty miles away
from the Siberian explosion

After the terrible flash—which, Father Kleinsorge later realized, reminded him of something he had read as a boy about a large meteor colliding with the earth—he had time (since he was 1,400 yards from the center) for one thought: A bomb has fallen directly on us. Then, for a few seconds or minutes, he went out of his mind.

Survivor of Hiroshima explosion, 1945,
from John Hersey's *Hiroshima*

THE FIRE CAME BY

Preface

This book tells the story of an explosion that occurred almost seventy years ago in the remote wilds of central Siberia. It is essentially a mystery story, for though the blast was among the most powerful ever occurring before or since on earth, its exact cause is not known; it remains today one of the greatest scientific riddles of all time. In telling the story, we have decided to let the mystery unfold chronologically, to relate the events as they happened and to allow the people involved to speak for themselves, particularly the eyewitnesses whose words provide a graphic record not only of the event itself but of their terror in the face of an inexplicable catastrophe.

The book is also a chronicle of history, covering five decades of expeditions to the Tunguska region in Siberia and all of the major theories offered to explain the blast. The evolution of modern science plays a crucial role in the narrative, as each generation of investigators attempts to cope, within the limits of its knowledge, with the extraordinary facts of the event. The members of the small, ill-equipped early expeditions were like detectives working with hopelessly primitive tools to solve a crime that was beyond their understanding; yet even the most recent expeditions, large teams of experts aided by the most sophisticated equipment, have been baffled by the complexity of the evidence and the tremendous magnitude of the blast.

The unparallelled nature of the Siberian explosion is illustrated by comparison with the most violent natural explosions of history, such as those that destroyed Thera, Pompeii, or Krakatoa. The advent of the atomic age provides man-made comparisons, but scientists find that even the awesome nuclear fire unleashed by the bomb dropped from the B-29 *Enola Gay* in 1945 over Hiroshima is dwarfed by the force of the 1908 blast. Only the later thermonuclear tests conducted by the United States and the Soviet Union can approach it in sheer size and scope.

The search for the identity of the mysterious "cosmic body" that caused the Siberian explosion involves mankind's oldest science, astronomy, as well as basic issues of cosmology dealing with what astronomer Fred Hoyle has called the "universal game that is going on around us, the game of which we are perhaps a tiny part." At the beginning of this century, at the time of the explosion, the extent of the universe was believed to be our galaxy; and the study of celestial objects such as meteorites and comets was then in its infancy. By the 1970s, however, astrophysics and space technology provide an expanded vision of a turbulent and infinite metagalactic universe full of bizarre objects like particles of "antimatter" and collapsed stars called "black holes" which defy the established laws of physics. Each of these newly discovered phenomena is examined as a possible answer to the Siberian puzzle.

In Russia the event has acquired an almost legendary status. In addition to hundreds of scientific papers, the great blast has inspired countless stories, poems, films, and television programs. A bibliography compiled in 1969 by a Soviet periodical listed more than a thousand items published about the catastrophe of 1908.

In our research we have concentrated primarily on original Russian scientific sources, which have the most reliable, accurate accounts of the event. With the help of our translator, John W. Atwell, we have been able to include new material and information not previously available in English. Although for the sake of readability, we have not used footnotes in the text, a full bibliography of references appears at the end of the book. Also included are two primary documents, translated especially for this volume, containing important eyewitness descriptions of the event.

Finally, like all mysteries, the Siberian explosion beckons to be answered and put to rest. In the last chapters we present the latest findings of Soviet scientists and then, after providing a cosmological justification for our conclusion, attempt a reconstruction of the morning that the fire came by.

Thomas Atkins
John Baxter

Bang!

Once upon a time there was no planetary system, just a cloud of dust and gas moving around a central region that was collapsing and becoming the Sun. Little by little, however, the dust and gas gathered into larger pieces, which slowly collected into world-sized bodies.

As the pieces grew larger, their gravitational fields grew more intense and the mutual collisions became harder and more forceful. Finally, when most of the material had collected into a large, solid sphere, the last large pieces that joined that sphere came down with a crash that kicked up vast craters.

The Moon is covered with such craters, as we have known for nearly four centuries. In the last decade, we have discovered that Mercury is covered with them, too, as is Mars and even the two tiny satellites of Mars. The first very crude close-up pictures of Ganymede, the largest satellite of Jupiter, indicate the presence of craters.

Venus, under its obscuring clouds, probably possesses them, too.

But what about Earth? Was our world not formed by gradually strengthening collisions as the others were? Should there not be the marks of those last impacts upon it?

There would be, if Earth weren't different from the others. Unlike the other worlds I've mentioned, it has an ocean which can absorb an impact and remain unmarked. Rain, rivers, wind, and, most of all, the busy bustle of living things act to erase any impact marks that form on land.

Even so, there are some signs. There are worn-down, waterlogged circular patches visible from the air here and there on Earth, and these may be the remains of ancient craters. In Arizona, about 30 kilometers west of the town of Winslow, there is an actual crater which must have been gouged out by the impact of a thudding meteorite some thousands of tonnes in weight. The hole is 1,200 meters wide and 800 meters deep and is called "Meteor Crater."

Why is Meteor Crater still there? First, because it is recent. The meteorite that formed it may have fallen as recently as ten thousand years ago. Secondly, it fell in a desert area where the action of water and life is minimal.

And has anything happened since Meteor Crater was formed?

Impacts are not as likely to take place now, as in primordial days. After all, the space near the worlds that now exist has been swept comparatively clean by the catastrophes of the past. Every mark on the Moon is the result of a fragment of rock we don't have to worry about any more.

Yet space is not *entirely* clean. There are known objects up to 25 kilometers across that can approach within 25 million kilometers of Earth. These won't actually collide with us, but there must be large numbers of minor bodies, much smaller, that come much closer. These small bodies are not likely to be seen at all until they enter the atmosphere. And though they're small compared to the monsters that were common in the past, they're still large enough to do enormous damage on the human scale.

The potential is there, then—but has anything *happened*?

We might speculate about some of the legendary catastrophes of the past. Did a meteorite speed into the Persian Gulf and splash half its contents of water into the Tigris-Euphrates valley—and was that the origin of the tale of Noah's flood? Was it a meteorite, or a group of them, that fell on Sodom and Gomorrah and destroyed them with fire from heaven?

We can't say. We don't know. We may never know.

But is there anything that happened that is *more* than speculation?—Yes, there is.

Once, and only once, in known history was there a clear and documented event that looked as though a large meteorite had fallen on Earth. It took place only seven decades ago, in 1908, in Siberia.

It was an amazing fall. On the one hand, it did enormous damage, for it fell in a forest and knocked down every tree for scores of miles in every direction. On the other hand, it did very little damage, for it killed not one human being.

Consider how unusual that had to be.

Seventy percent of the Earth's surface is water. If that fall had taken place anywhere in the ocean, tsunamis (so-called "tidal waves") would have washed the nearer shores and done much damage. Another 10 per cent of the Earth's surface is covered by permanent ice. If the fall had taken place there, enough melting might just barely have come about to cause the slippage of large quantities of ice into the ocean, bringing about catastrophic changes in Earth's sea level and climate.

At least 15 per cent of what is left of Earth's surface is populated, more or less thickly, with human beings; and is littered, more or less thoroughly, with the products of their civilization. If the fall had taken place there, anywhere from hundreds to millions of people would have been killed and anywhere from thousands to billions of dollars of damage would have been inflicted. The fall would have completely wiped out any city it had struck.

Perhaps not more than 5 per cent of the surface of the Earth could have received that blow without any damage at all being done to human life and property. And, with the odds twenty to one against it, that fall took place safely (from the human standpoint).

By the same token, though, the place in which the fall occurred was inaccessible (else it would have been populated) and it was years before the vicinity could be examined. It was only then that the *real* mystery began. I won't tell you what that was, because that is what the book is about, and the book tells it better than I can.

Baxter and Atkins tell the story with great skill; doing it methodically, in order and in detail, taking every factor into account, and going about the process, very rationally, of making an absorbing suspense tale out of it. And in the end, they come up with a dramatic solution indeed.

You may not accept the authors' solution, for it is reached by elimination and by uncertain report, rather than by direct evidence (which, by the nature of things, cannot exist), and such a line of argument is always shaky. But think about it—

In the light of the authors' solution, consider that the fall managed to find a one-in-twenty place where it would do no damage, almost as though someone was humanely trying to avoid—

But read the book! I loved every page of it!

Isaac Asimov

One

THE EXPLOSION

High above the Indian Ocean, a huge object hurtling from space pierces the earth's atmospheric shell. In the almost airless upper altitudes, there is no sound, barely any friction; unimpeded, it races toward the earth.

In a long sloping trajectory, it rockets northwards at supersonic velocity across the Asian mainland, high over the Himalayan peaks. Drawn lower by gravity, it hits the thicker strata of the globe's atmosphere. Intense frictional heat begins to build up.

It first flashes into the sight of man over western China in the dawn of June 30, 1908. Caravans winding through the Gobi Desert halt and look in awe at a fireball blazing across the sky. It disappears over the border of Mongolia. Plunging into the denser air layers, it glows with the heat of 5,000 degrees Fahrenheit, brighter even than the thin morning sun.

In central Russia, a deafening roar terrifies the inhabitants of small towns and villages, the only settlements in this remote and deserted area. A powerful ballistic wave pushed before the descending object strikes the ground. Trees are leveled, nomad huts blown down, men and animals scattered like specks of dust.

At 7:17 A.M. the Central Siberian Plateau near the Stony Tunguska River, a sparsely populated, desolate region of peat bogs and pine forests, shudders under the impact of a cataclysmic explosion.

The detonation is of such violent force that the seismographic center at Irkutsk, 550 miles to the south, registers tremors of earthquake proportions. Vibrations travel 3,000 miles through the ground to other stations in Moscow and the capital of the tsarist empire, St. Petersburg; and the earthquake observatory at Jena, Germany, 3,240 miles away, records strong seismic shocks. Even as far away as Washington and Java seismographs are activated by the immense blast.

Instantly, a gigantic "pillar of fire" flares up into the clear blue sky, ascending to such height that the blinding column is visible above the horizon to startled Siberians in towns several hundred miles away; then the air is wracked by a series of thunderous claps that can be heard for more than 500 miles. The noise is so great that some herdsmen closer to the blast are deafened; others are thrown into a state of dazed shock that renders them speechless.

Simultaneously with the brilliant fire in the sky, a searing thermal current sweeps across the hilly taiga, or northern woods, scorching the tall conifers and igniting fires that will continue to burn for days. Stunned citizens 40 miles away in the trading post of Vanavara shield their faces from the fierce heat drafts. Seconds later, a shock wave generated by the blast rips through the small village, gouging up pieces of sod, collapsing ceilings, shattering windows, and flinging people into the air.

At a distance of 375 miles to the south-southwest, hurricanelike gusts rattle doors, windows, and lamps in Kansk, a station town on the newly completed Trans-Siberian Railway. Within minutes two additional waves of shock strike the town. People working nearby on rafts are hurled into the river, while farther south, horses stumble and fall to the ground.

Near Kansk, aboard the Trans-Siberian Express, passengers are frightened by loud bursts of noise and almost jolted out of their seats. The train is jarred and shakes wildly on its tracks. As the startled engineer sees the rails ahead vibrating, he quickly brings the train to a screeching halt. Eventually, when the ground upheavals subside, the locomotive proceeds to the nearest station where the engineer and a station agent make sure that the travelers are unharmed and inspect the whole train to see whether any of the freight has been damaged.

As dark masses of thick clouds rise to an altitude of more than 12 miles above the Tunguska region, the entire area is showered by an ominous "black rain," the result of sudden air condensation and the fountain of dirt particles and debris sucked up into the swirling vortex of the explosion. Intermittent rumblings of thunder, resembling heavy artillery, reverberate throughout central Russia.

Far across the continent to the west, in St. Petersburg, no one hears of the blast, nor will anybody hear of it for many years. The record of seismic shocks, when they are noticed, will at first be mistaken for earthquake tremors. In 1908 most Russians have their minds on other problems, for their country is gripped by social unrest and political tension that eventually will erupt in the Revolution. From his imperial capital, an elegant city whose cultural closeness to Europe emphasizes the government's detachment from its own citizens' troubles, Tsar Nicholas II maintains a precarious hold on a nation divided and, in many cases, desperate.

To avert the threat of revolution in 1905, Nicholas reluctantly agreed to the formation of Russia's first parliament, the Duma; and the elections of 1907 had brought in many Social Democrats and veterans of earlier revolutions against tsarist autocracy. For the first time, Russia has labor unions—though they lack the power to strike—and a moderately free press. But the universities in particular remain dissatisfied; among the men whose revolutionary activities at this time will land them in jail is a young forestry student named Leonid Kulik, who is to become a central figure in the solution of the Tunguska mystery.

To many students, the new Duma seems as uninterested in Russia's problems as are the councils of the Tsar. Throughout June and July of 1908, the news in St. Petersburg is not of catastrophes such as the Tunguska explosion—which is mentioned only in local Siberian papers—but of events like a duel fought between two Duma members over a matter of political disagreement. The relocation of Russia's capital from St. Petersburg to Moscow does not take place until 1918; meanwhile, the government remains isolated and indifferent to the disasters and mysteries of its vast, unknown country in the east.

The appalling misery of this oppressed land as well as its sense of mystery and promise are evoked by Nikolai Gogol in his tragicomic masterpiece *Dead Souls*. Describing a trip taken by his hero Chichikov through the Russian steppes, he writes: "I see you: a country of dinginess, and bleakness and dispersal. . . . So what is the incomprehensible secret force driving me toward you? Why do I constantly hear the echo of your mournful song as it is carried from sea to sea through your entire expanse? . . . And since you are without end yourself, is it not within you that a boundless thought will be born?"

In 1908, across the Pacific, in the United States, where Teddy Roosevelt is serving the final year of his second term as President, few people pay attention to the upheavals occurring in the Tsar's distant country. No word of the Siberian explosion has yet reached the United States. Instead, Americans are watching the appearance of several new machines that soon will revolutionize transportation, communication, and warfare throughout the world. Businessman Henry Ford has just released his Model T, the first mass-produced automobile; within less than two decades twenty-five million of these vehicles will be swarming across the American landscape.

The United States War Department, after dropping his initial letter in the "crank" file, finally decides to give inventor Orville Wright a contract to build the first military airplane. In August his brother Wilbur amazes the French press by doing a figure eight in his flying machine and staying aloft for 107 seconds at Hunaudières racecourse; and by September, at Fort Myer, Virginia, Orville manages to keep his machine in the air for one hour and five minutes. Two years later the United States Government conducts the first experiments in aerial bombing.

Around the world, scientists armed with the latest inventions of nineteenth-century technology are struggling to interpret the universe to the newly born twentieth century. Astronomers, observing the solar system and our galaxy through increasingly powerful telescopes, have gathered fresh and often puzzling information about the behavior of remote stars and other cosmic phenomena. In 1908 the cosmos is still regarded as basically stable and unchanging; its "edge" is thought to be the Milky Way, which has not yet been fully mapped.

Meteorological instruments, though still relatively primitive, are used to explore the intricate relationship between solar flares and the earth's magnetic fields, but without a solid understanding of nuclear physics such readings cannot be fully analyzed. At this time Robert Millikan is doing his ground-breaking work on X-rays, and the "ether" theory of space has just been overturned. Albert Einstein will soon radically change scientific notions of space, time, and matter with his general relativity theory. The neutron has not yet been discovered by James Chadwick. The Nobel Prize in chemistry for 1908 is awarded to Ernest Rutherford for his pioneer studies of radioactivity; he will shortly formulate his atomic nucleus idea that is to become one of the fundamental concepts of modern physics.

The year 1908, in retrospect, seems to be a time of strange and inexplicable events, a period when people claim to see peculiar moving lights at night or captains report "magnetic clouds" descending on their ships. "Alien airships" are seen racing at fantastic speeds over the United States and Europe. That summer a "200-foot-long sea serpent" rises from the Gulf of Mexico off the Yucatan Peninsula and is described in great detail by passengers on a steamship.

Scientists watching the sky in 1908 are prepared for new occurrences, yet poorly equipped to identify or interpret them—particularly one as unprecedented and complex as the event near the Stony Tunguska River.

Though news of the explosion has not appeared in the American or European press and scientists in the West as yet know nothing of the event, the papers during the summer of 1908 are full of speculation about the unusual meteorological phenomena and magnetic disturbances that follow the devastating blast. A report from Berlin in the New York *Times* of July 3 attempts to explain the bizarre colors seen recently as the Northern Lights:

> Remarkable lights were observed in the northern heavens on Tuesday and Wednesday nights, the bright diffused white and yellow illumination continuing through the night until it disappears at dawn. Director Archenbold of the Treptow Observatory says that because of the phenomenon's particular brilliancy he thinks it may be connected with important changes in the sun's surface, causing electrical discharges. Director Archenbold, however, mentions a somewhat similar phenomena in 1883, which was directly traceable to an outbreak of the Krakatoa volcano in the Strait of Sundy [Sunda]. Reports from Copenhagen and Königsberg told of the same great lights being visible in these cities, and it is presumed that they were visible throughout Northern Europe.

But the 1883 eruption in the East Indies is similar only in its vast size and strength; the 1908 Siberian explosion is not volcanic, and the time scale and physical characteristics of the phenomena resulting from the event are utterly different—and ultimately far more frightening and mysterious—than those that followed Krakatoa.

Unusual atmospheric charges prompt speculation in England the evening after the explosion. "There was a slight, but plainly marked, disturbance of the magnets on Tuesday night," states an editorial in the *Times* of London; but this magnetic disturbance is at first mistakenly associated with "disturbances in the sun's prominences" rather than with the Siberian catastrophe, which has not been reported in British newspapers at the time.

Within five hours of the blast, turbulent air waves travel west beyond the North Sea, causing strong oscillations at meteorological stations in England. During a span of twenty minutes, sudden fluctuations in atmospheric pressure are detected by recently invented self-recording barographs at six stations between Cambridge, 50 miles north of London, and Petersfield, 55 miles south. Baffled weather researchers assume that a large atmospheric disturbance has occurred somewhere in the world; but not until two decades later, after the first news of the devastation in the Tunguska region finally reaches the English press, do they discover that their 1908 barographic records correspond with the astonishing Russian explosion of that same year and that its air waves circled the globe twice.

But perhaps even more striking than the unexplained effects on the earth's magnetic fields and the powerful air waves associated with the blast is the appearance at high altitudes of massive luminous "silvery clouds" blanketing Siberia and northern Europe. The light is so intense during the next few nights that in some places it is possible to take photographs at midnight and ships can be seen clearly for miles out at sea. A Russian scientist describes the thick layer of glowing clouds as lit up by "some kind of yellowish-green light that sometimes changed to a rosy hue." He adds, "It was the first time I had seen such a phenomenon."

Extraordinary dust clouds and eerie nocturnal displays, moreover, are observed for weeks across the Continent as far south as Spain. On June 30, the day of the explosion, a scientist in Holland sees "an undulating mass" passing across the northwest horizon. "It was not cloud," he states, "for the blue sky itself seemed to undulate." That same evening an astronomer in Heidelberg finds that when he tries to photograph the stars, his plates are fogged by the abnormal luminosity in the sky. After sunset in Antwerp the northern horizon appears to be on fire.

In 1930, in the *Royal Meteorological Society Quarterly Journal*, Spenser Russell gives an account of the odd colors he observed over England in 1908 on the nights of June 30 and July 1:

> A strong orange-yellow light became visible in the north and northeast . . . causing an undue prolongation of twilight lasting to daybreak on July 1st, when the eastern sky was an intense green to yellow-gold hue. . . . The entire northern sky on these two nights, from the horizon to an altitude of 40°, was of a suffused red hue varying from pink to an intense crimson. There was a complete absence of scintillation or flickering, and no tendency for the formation of streamers, or a luminous arch, characteristic of auroral phenomena. . . . Twilight on both of these nights was prolonged to daybreak, and there was no real darkness. . . . The phenomenon was reported from various places in the United Kingdom and on the Continent, from Copenhagen, Königsberg, Berlin, and Vienna.

According to the London *Times* of July 4, 1908, "The remarkable ruddy glows which have been seen on many nights lately have attracted much attention, and have been seen over an area extending as far as Berlin." The cause is assigned to "some condition of the atmosphere," such as occurred after Krakatoa, although "no volcanic outburst of abnormal violence has been reported lately." The *Times* notes that the recent "abnormal" glows appear in the sky only *after* the fading of twilight: the sky grows partially dark and then brightens again with "deep, lurid color."

On July 5, in a New York *Times* report from Britain entitled "Like Dawn at Midnight," a correspondent writes:

> Following sunsets of exceptional beauty and twilight effects remarkable even in England, the northern sky at midnight became light blue, as if the dawn were breaking, and the clouds were touched with pink in so marked a fashion that police headquarters was rung up by several people who believed that a big fire was raging in the north of London.

In the London suburbs citizens are drawn into the streets to view the frightening cosmic phenomenon. A woman in Huntingdon, north of London, alarmed by the vivid night lights, writes to the London *Times* that shortly after midnight on July 1 the sky was so bright that "it was possible to read large print indoors, and the hands of the clock in my room were quite distinct. An hour later, at about 1:30 A.M., the room was quite light as if it had been day; the light in the sky was then more dispersed and was a fainter yellow." She concludes, "I have never at any time seen anything the least like this in England, and it would be interesting if anyone would explain the cause of so unusual a sight."

But no satisfactory explanation is offered to the woman's question. The silence about the nature of the incredible nightly spectacles seen over Europe and the riddle of the great Siberian explosion is not broken for more than a decade.

Two

AN ENORMOUS SILENCE

Human history is marked indelibly with the destruction caused by great explosions.

In about 1500 B.C. most of the island of Thera (later called Santorini) in the Aegean, 70 miles north of Crete, disappeared in a blast which reduced a fertile kingdom to little more than a splintered shell. Today, one can see the crescent rim of the dead volcano and two blackened, still-active islets in the center of a 16-mile-wide lagoon where once stood an important center of Aegean culture. Sheer cliffs plunging a thousand feet into the dark water, their sides striped with the brilliant colors of blistered rock, attest to the force of the blast.

Tidal waves from the explosion swept across 70 miles of sea to Crete, rolling over the palaces and temples, crushing the fragile frescoes and pillared colonnades of a society among the most enlightened in the world. Recent excavations on Thera have uncovered wall paintings of unexampled color and delicacy, as well as signs of a thriving and sophisticated community—discoveries which lend weight to the theory that in the sudden destruction of Thera and the related Cretan empire may lie the beginnings of the legend of Atlantis.

No similar catastrophe affected history until A.D. 62, when an earthquake caused by the eruption of Vesuvius toppled much of the towns of Pompeii and Herculaneum near Naples. Many of the houses had not been rebuilt seventeen years later when, on August 24, A.D. 79, Vesuvius erupted once more; in two days, both towns were inundated with hot ash, small stones, and lava as deep in some places as 23 feet. In Herculaneum, water mixed with this debris to create an impermeable plasterlike covering which has been gradually cleared by archaeologists in the last three centuries to reveal, frozen in time, the life of a busy Roman town.

The Vesuvius blast was dwarfed by what experts generally agreed to be the greatest natural explosion of modern times, the destruction of the volcanic island of Krakatoa between Java and Sumatra. Krakatoa, which had erupted intermittently for centuries, shuddered into life again in May of 1883. More eruptions were felt during June and August. On the early afternoon of August 26, the volcano began to explode, eventually discharging a 17-mile-high cloud of dark ash; and at 10 A.M. on the following morning the whole island was shaken by a cataclysmic blast. Ash fountained 50 miles into the sky. Air waves from the explosion registered all over the world, and the sound was heard 2,200 miles away in Australia. Five cubic miles of rock disappeared, some falling back as stones and ash on the surrounding islands and clogging the waters so badly that nearby ships were unable to move through the debris.

Tidal waves more than a hundred feet high smashed into the Java and Sumatra coasts, killing 36,000 people; the same waves registered on the coasts of South America and Hawaii. For several days, darkness masked the area around the volcano. Thrown into the higher atmosphere by the enormous force of the blast, tons of dust were carried around the world a number of times by the high-velocity winds—55 to 110 miles per hour—of the stratosphere between 12 and 25 miles above the earth. The dust so saturated the upper air that sunsets all over the world carried a noticeable bloody tinge for more than twelve months and interference with solar radiation lowered the world temperature 0.5 degrees Fahrenheit for many years.

For all their enormous size, destructive capability, and long-term effects, none of these explosions could surpass the sheer force of the Tunguska blast in 1908. Its size, as registered on seismographs and barographs in Europe, deserves comparison only with the greater man-made explosions of the atomic age; even the Hiroshima blast and the nuclear tests of the early 1950s are dwarfed by it. Authorities in the U.S.S.R., Britain, and the U.S.A. agree that the estimated energy output of the Siberian blast—10^{23} ergs—would be comparable only with the explosion of the heaviest hydrogen bombs.

Today, the failure at the time to investigate the Tunguska explosion seems at first incredible; a cataclysm greater than any the world had ever known went without serious inquiry or comment for thirteen years and even then was discovered only by the most remarkable series of coincidental events and the stubborn efforts of one man. But, for numerous reasons, the vast silence of 1908 to 1921 is understandable.

Most catastrophes in human history have gone, by modern standards of communication, largely unrecorded; and the fact that the Tunguska explosion went unexplored for such a long period is not unusual. Chiefly because they were volcanic, the explosions of Thera, Vesuvius, and Krakatoa happen to be exceptionally well documented.

Volcanoes attract attention. In general, they herald their major eruptions with a fanfare of minor tremors, emissions of smoke, and showers of ash and leave behind, as in the case of all three major explosions, a wealth of geological evidence which, even on the island of Thera, can still be read thousands of years later. In addition, because volcanic ash provides the basis for rich and fertile soil, most volcanoes are surrounded by thriving communities and thus a generous corps of observers. Krakatoa had not been still for five years after the 1883 eruption, in fact, before a community was again springing up on its slopes. Thera today supports 9,000 people and has thriving vineyards.

By contrast, some disasters of even greater force, because of their special nature, lack even a fraction of the evidence of Krakatoa or Thera. Many aspects of the Hiroshima and Nagasaki atomic explosions remain obscure because of a lack of trained witnesses, while, as William Manchester points out in *The Glory and the Dream*, the details of the damage wrought by the great hurricane which shattered Long Island and New England in September, 1938, leaving about 700 dead and 1,754 injured, went largely unreported, the news being partly eclipsed by the Munich crisis and Hitler's invasion of Czechoslovakia.

Meteorite explosions, among which the Tunguska blast was initially categorized, have proved even less easy to document. Without giving any of the warning signs shown by a volcano, meteorites come rushing in at cosmic velocity—as fast as 25 miles per second—and, should they happen to reach the earth, it is often in a remote area; examples of meteorites injuring people or property in populated areas are extremely rare. Evidence of their trajectory and impact is often confused or inaccurate; and if they fell in the distant past, records may be dim or totaly nonexistent. The collisions which created the huge craters in Arizona and Quebec occurred from ten to fifty thousand years ago, and the actual impact holes remain as the primary evidence.

In the case of the Tunguska blast, its discovery and documentation were complicated by the nature of the terrain where the event occurred. The very name "Siberia" derives from a Tartar word meaning "sleeping land." The vast central region lying east of the Yenisei River was, and still is in many ways, one of the most remote places on the face of the earth. Had the area been picked out as the location for such an event, it could not have been more secret. Significantly, when the Soviet government chose a place to explode its first atomic bomb in 1946, it was to Siberia and the frozen tundra above the Arctic Circle that it looked for a test site.

Kansk, one of the main railroad towns nearest to the explosion, is 2,500 miles from Moscow and 3,000 from St. Petersburg, the national capital in 1908—a distance as great as from New York City to Los Angeles. Before the building of the Trans-Siberian Railway, travelers on foot took as long as a year to make the journey from St. Petersburg to Irkutsk; and for more Russians at the turn of the century Siberia remained an area as foreign and distant as the moon. Besides the local farmers and fishermen, many of its inhabitants were political prisoners or their descendants; generations of dissidents had been shipped to centers like Irkutsk and Krasnoyarsk, forced to slave in the mines and forests, then had been abandoned there by a government which knew they had no choice but to settle and farm. Both Lenin and Trotsky served sentences in Siberia; in 1908 the political prisoners in the area almost certainly included Josef Stalin, who was sentenced to six different terms between 1903 and 1913.

From April to July of 1890, the Russian writer Anton Chekhov made an incredible trip from Moscow, chiefly by horse carriage and river boat, across the entire breadth of Siberia to Sakhalin, an island off the Pacific coast used by the Tsar as a penal colony. In his letters Chekhov provides vivid impressions of the Siberian climate, landscape, and people. In May, near Tomsk he wrote:

> Spring hasn't yet arrived here. There is absolutely no greenery, the forests are bare, the snow has not all melted and lusterless ice sheathes the lakes. On the ninth of May, St. Nicholas Day, there was a frost, and today, the fourteenth, we had a snowfall of about three inches. Only the ducks hint of spring. . . . I have never in my

> life seen such a superabundance of ducks. . . . You can hear the wild geese honking. . . . Often files of cranes and swans head our way. . . . Well, you go on and on. Road signs flash by, ponds, little birch groves. . . . Now we drive past a group of new settlers, then a file of prisoners. . . . We've met tramps with pots on their backs; these gentlemen promenade all over the Siberian plain without hindrance. On occasion they will murder a poor old woman to obtain her skirt for leg puttees [wrappings] . . . but they won't touch people in vehicles.

Although Chekhov praised the Siberians in general as good, honest people, he confessed that he did not find the women attractive. "Siberian women and misses," he remarked, "are frozen fish. You'd have to be a walrus or seal to have an affair with them." After having journeyed the 800 miles from Tomsk to Irkutsk, he commented: "The Siberian highways have their scurvy little stations, like the Ukraine. They pop up every 20 or 25 miles. You drive at night, on and on, until you feel giddy and ill, but you keep on going, and if you venture to ask the driver how many miles it is to the next station, he invariably says not less than twelve." On the way to Irkutsk Chekhov had stopped at Krasnoyarsk, caught a boat across the Yenisei, which he described as the "fierce and mighty warrior" of rivers, and then admired the mysterious and seemingly boundless taiga of central Siberia.

Though it covers a vast area, larger than Alaska, central Siberia has a relatively sparse population of only a few hundred thousand. Its migrating tribesmen, the Tungus (a Mongoloid people, later renamed Evenki by the Soviets), traversed only the areas best for trapping, so huge sections of the taiga were still unexplored. An American prospector of the early 1900s wrote of the difficulties of the Siberian taiga: "I had a terrible journey through forests and over mountains, where rain fell incessantly and I nearly died through exhaustion. The valleys, hillsides, in fact everywhere we went, was covered with bog made by the falling pine-needles through countless ages. The horses plunged on for miles in this stuff up to their knees."

Distances here were given in terms of "summer" or "winter," those for winter being much shorter, since the snow enabled sleds to cross areas impassable during the brief thaw. Roads in the European sense were unknown; the trappers followed tracks worn by generations of foxes and hares and seldom imitated the foolish foreigners who, searching for Siberia's enormous gold and mineral wealth, struck out into the wilderness. In such places, distance and time had little relevance, as investigators found when they tried to piece together eyewitness reports of the 1908 event—another factor contributing to the silence that surrounded the explosion.

Within central Siberia, the region of the Podkamennaya Tunguska, or Stony Tunguska, River is even more impenetrable. One of the many streams that drain north into the huge Yenisei, which in turn feeds the Arctic Ocean, the Stony Tunguska rises in a mountainous wilderness of rolling taiga spotted with huge swamps. The taiga in this region is no ordinary woodland but, as Yuri Semyonov points out in his study of Siberia, a "vast and sinister" primeval forest in which "the weak and imprudent often perish" in its trackless depths and pathless bogs where "everything below is decayed and rotten, and everything above withered, where only the corpses of the huge trunks slowly moulder away in the brackish water."

Its chief inhabitants, the nomadic Tungus, survived mainly on their reindeer herds and on hunting other animals—such as bear, fox, and sable—for their pelts. The few trading posts, situated along the riverbanks, were tiny and primitive. The Tunguska blast occurred during the height of the short summer, when the soil of the taiga melts to a swampy slush and the air is infested with ferocious Siberian mosquitos. The winter frosts of the area, which lies about 300 miles below the Arctic Circle, can be so terrible that, according to Semyonov, birds drop from the air as if dead and freeze immediately unless taken inside for warmth.

In addition to its remoteness and inhospitable climate, central Siberia had an atmosphere of ancient mystery and an accumulation of inexplicable phenomena. For centuries, the country around the Stony Tunguska had been gathering its share of strange stories and legends: weird species of fish existing in the depths of Lake Baikal, giant subterranean rats living beneath the ground, monstrous mammals encased in icy tombs. The explosion of 1908, just one more in a series of curiosities, must have seemed to some natives, and later to people in Moscow and St. Petersburg, hardly worth the trouble of comment.

This is the home of the "Siberian disease," or "Arctic hysteria." Historian Emil Lengyel wrote: "It is mimicry mania. The victim repeats whatever he hears or sees. Sometimes he repeats words he does not even understand or repeats animal sounds, such as barking. This malady also takes other forms. At village dances, it sometimes happens that the rhythm becomes too strong to resist and the dancers cannot stop. The hysteria becomes infectious and even old people are caught in the swirling emotion and dance until exhaustion makes them collapse." Political prisoners often succumbed to the disease; a German officer sent to Siberia as a prisoner of war in 1918 repeated the word "dismal" constantly in a dozen languages, while another mumbled the phrase "Life is death and death is life" over and over again until taken to the camp psychopathic ward.

To the southeast of the Tunguska region lies Lake Baikal, one of the most unusual bodies of water in the world. Lying across a fault in the earth's crust, the Lake Baikal area has been riven by earthquakes for millennia; the large number of shocks in the late nineteenth century obviously influenced scientists at first to class the 1908 explosion as seismic. Millions of years ago the plates on which this section of Asia is balanced ground against one another, throwing up massive mountain ranges and forming a fissure, as the plates resolidified, more than a thousand miles long, thirty to forty miles wide, and several miles deep. Normally, such gaps are filled quickly with molten material, but this one failed to close up completely and over thousands of years a 400-mile-long rift filled with fresh water to become Baikal.

Departing from Irkutsk for the final stretch of his trip to Sakhalin, Chekhov crossed the "astonishing" Lake Baikal. "It is with good cause that the Siberians entitle it not a lake, but a sea," he wrote in a letter of June 20, 1890. "The water is unusually limpid, so that you can see through it as you do through air; its color is tenderly turquoise, pleasant to the eye. The shores are mountainous and wooded; all about are impenetrable, sunless thickets. There is an abundance of bears, sable, wild goats, and all kinds of wild game, which occupies itself in existing in the taiga and making meals of one another."

Baikal is the largest body of fresh water in the world, a mile deep in some places, and harboring some of the oddest creatures known to science; of the 1,800 plants and animals in its waters, a thousand exist nowhere else on earth. Seals and fish normally found only in salt water live comfortably there, hundreds of miles from the nearest ocean; and Baikal is the only home of fish like the *golomyanka*, which lives at depths that can crush a steel tube. "It is hardly ever seen alive," remarked the Western traveler Bassett Digby in the late 1920s. "The shock of the inshore breakers seems to kill it. This little fish is only a few inches long, at the most. Its flesh is so soft and oily that it readily melts like butter. Even the heat of the sun is sufficient to reduce the dead fish to a mere head, backbone and flabby strip of thin skin, lying in a pool of oil. The natives patrol the shore for it after wild weather, sometimes melting it down for lamp fuel and sometimes eating it." According to the lore about the creature, "Quantities of golomyankas are sometimes thrown up after volcanic disturbances. The Tungus natives repeat an old legend that has been passed down by bygone generations to the effect that storms in the mountains rush down through subterranean abysses leading under the lake and blow up these fishes from the great depths where they seem to dwell."

Baikal encourages such legends. Even when the surface is so deeply frozen that heavy vehicles can travel across without risk, waves race under the ice and shake the shore. In the thaw, the sound of the ice splitting can be heard ricocheting around the mountains like artillery fire. Scientists have no explanation for the lake's extraordinary variations in depth; in the summer of 1818 the surface suddenly rose 6 feet and remained at that level for some time, though no blockage was found of any river that drained the lake.

Subterranean horrors loomed large in the mythology of the Tungus, who explained the frequent earth tremors in their region as signs of herds of giant rats that burrowed constantly under the earth, making tunnels in which their tramping feet set up the rumblings of earthquakes. They also described strange races of men inhabiting the unexplored areas of Siberia. One was said to have mouths in the top of their heads, others thick hair all over their bodies. Another group they said hibernated for two months, totally oblivious to every stimulus.

Scientists readily explained these "monsters" as distortions of tribes seen by the Tungus only occasionally in this huge and remote area. Those covered in hair merely wore tight-fitting fur coats, unlike the Tungus's reindeer hide covering. "Eating through the top of the head" was a distortion of the antics necessary for a man in a heavy coat and hood to get food into his mouth. And to the Tungus, nomads with few possessions larger or heavier than iron axes, the carved wooden or ivory statues of more settled tribes looked like men frozen into a catalepsy, both unable to feel and to see when they were brought out of winter hiding.

Travelers to the Tungus' region were also inclined to explain away the legend of the giant earthquake rats as sheer fantasy until a young surveyor named Benkendorf, hired to map out areas of eastern Siberia for the tsarist government, reported some details of his trip in 1846. Mapping an area on the Lena River in an unseasonably hot summer he noted, after pitching camp, that "the stream was tearing away the soft sodden bank like chaff, so that it was dangerous to go near the brink. In a lull in the conversation we heard, under our feet, a sudden gurgling and movement in the water under the bank. One of our men gave a shout and pointed to a singular shapeless mass which was rising and falling in the swirling stream. . . a huge black horrible mass bobbed up out of the water. We beheld a colossal elephant's head, armed with mighty tusks, its long trunk waving uncannily in the water, as though seeking something it had lost. Breathless with astonishment, I beheld the monster hardly twelve feet away, with the white of his half-open eyes showing. 'A mammoth! A mammoth!' someone shouted."

Scientists had known that the hairy, elephantlike mammoths roamed the central taiga and polar regions of Siberia into relatively recent times. Many tribes had objects carved from mammoth ivory, but few preserved bodies had been found. The early discovery of these bodies, which rotted too quickly to be saved, led to a rash of similar excavations and to an explanation of the old legend of the giant rats. Many mammoth bodies were found when riverbanks crumbled or landslips revealed frozen ground. With its tusks jutting out of the bank or its huge hairy feet exposed, it probably looked uncannily as if some huge burrowing animal had been frozen in the act of tearing its way out of the earth. Most Tungus would no doubt have fled in panic, bringing back stories of a ratlike snout and fangs, a hairy face, and wide burrowing paws.

Travelers' tales of mammoth hunting in Siberia built up a mythology around the turn of the century often as grotesque as that invented by the shamans of the Tungus. Reporters told of mammoths so well-preserved that their flesh was edible, of bodies wedged in crevasses with certain signs of recent life still visible—the food in the stomach undigested, the eyes still clear. Incredibly, most of the reports were completely true. A Russian expedition excavating a mammoth at Berezovka in 1901 wrote, "The flesh from under the shoulder, fibrous and marbled with fat, is dark red and looks as fresh as well-frozen beef or horse meat. It looked so appetizing that we wondered for some time whether we would not taste it. But no one would venture to take it into his mouth, and horseflesh was given the preference. The dogs ate whatever mammoth meat we threw them."

Though the mammoth had roamed over most of the world, it was known in temperate areas only through collections of miscellaneous bones, fossil remains, and prehistoric cave paintings. But in northern and large parts of central Siberia the underlying earth remains frozen at minus 4 degrees Centigrade for the entire year. Except for a few feet of the upper surface which softens in the summer thaw, the ground in much of central Siberia is as hard as concrete, a mixture of soil, rock, and water called *merzlota*, or permafrost, that has remained unchanged for millennia. Prehistoric animals like the mammoth, which fell into bogs or small streams in the thaw, were quickly covered in snow and buried still further by landslips or sedimentation. Captured by the permafrost, they remained much the same as when they died, the bacteria of decay numbed by the constant cold.

In his preface to Part I of *The Gulag Archipelago*, Alexsandr Solzhenitsyn recounts a news item describing the discovery, in a frozen subterranean stream in Siberia, of a variety of prehistoric fauna. "Whether fish or salamander," he writes, "these were preserved in so fresh a state, the scientific corrspondent reported, that those present immediately broke open the ice encasing the specimens and devoured them *with relish* on the spot."

Because of its bleakly cold climate and eternal frost, Siberia is one of the few places in the world where the effects of the 1908 explosion would remain unchanged for such a long time. Most of the scars of its dreadful firestorm did not heal quickly but were preserved, almost as if in a deep freeze.

In addition to causing a firestorm, the blast in the Tunguska region had radiated enough heat in a few seconds to melt the permafrost stratum to a great depth, causing swelling of large rivers and flooding. Before the early 1920s only a few of the bolder Tungus, at great risk to themselves it was later learned, had dared to enter this scarred region to see the damage. Indifference, misinformation, and falsely preconceived ideas, as well as the remoteness of the region, had helped to prevent any serious scientific investigation until then; but the official Soviet research body, the Academy of Sciences, was soon to take the first steps that would begin to change this situation.

Three

THE FIRST EXPEDITION

The search for an answer to the Tunguska mystery began in what was for Russia a period of total disorder.

In October of 1917, during World War I, revolution had broken out. Shortly afterward, Russia withdrew from the war and the revolutionary government signed a separate peace with Germany. The Western Allies, concerned that Russia might become a German satellite, sent an expeditionary force into northwest Russia. The Japanese landed a force on the Pacific coast of Siberia, and a large group of Czech prisoners of war marched toward Vladivostok in the hopes of eventually reaching their homeland. Many of the Czechs joined forces with Admiral Alexander Kolchak's White Army which was engaged in a bloody war of resistance to the Red Army in central Siberia. For three years much of Siberia was a battleground, littered with corpses.

"Siberia was thrown into anarchy," wrote Emil Lengyel. "Typhus broke out among the refugees; the hospitals could not accommodate them. They stampeded to the towns in the hope of finding relief. In one town of 70,000 inhabitants, struck by the refugee wave, about 200,000 perished. Thousands lay on the streets, at railway stations, along the roads. Hunger followed in the wake of the epidemic. Tens of thousands of monstrously blown-up human forms awaited merciful death in the town of Taiga alone."

Kolchak's men were pitiless in defeat. Heaps of bodies marked his passing as he retreated to Irkutsk, which he attempted to make his new capital. The Communist forces were little better. A refugee living near Krasnoyarsk, one of the larger towns in central Siberia, saw in the Yenisei River in winter of 1920 the result of the activities of the Cheka, or Soviet police, charged with wiping out Kalchak's men: "Hundreds of bodies with heads and hands cut off, with mutilated faces and bodies half burned, with broken skulls, floated and mingled with the blocks of ice, looking for their graves; or turning in the furious whirlpools among the jagged blocks, they were ground and torn to pieces into shapeless masses, which the river, nauseated with its task, vomited out upon the islands and projecting sand bars."

Split by civil war and threatened by foreign powers, the Soviet government had little time for scientific inquiry; it was not until 1921, with Kolchak executed, the Japanese and other foreign forces expelled, and order restored, that expeditions once again ventured into the interior of Russia and Siberia.

One of the first trips was mounted by the Academy of Sciences. Before the Trans-Siberian Railway was built the vast distances of Russia had hampered exploration. Now, with the war over and the railway back in operation, most areas of the remote back country were opened up; the Soviet government chose to exploit the situation by sending out not a prospecting or map-making team but one charged with tracking down and locating the many meteorite falls which had been recorded in Russia during and after the war years.

The government's incentive may have been financial.

An American group, the United States Smelting, Refining and Mining Exploration Company, had begun work at Meteor Crater (also known as Barringer Crater) near Winslow, Arizona, to exploit the largest meteoritic mass known to have hit the earth. Three quarters of a mile in diameter and 570 feet deep, Meteor Crater had intrigued prospectors for decades. In 1903 and 1908 the Standard Iron Company of Philadelphia sank shafts into the crater floor but found nothing. The new company, with a firmer grasp of meteorite ballistics, realized that the object had entered at an angle and drilled near the south rim. At 1,376 feet the bit jammed, after having brought up oxidized meteoritic iron for 200 feet; as the engineers tried to withdraw the drill, their cable broke, presumably wedged in the dense iron mass. Analyzing samples of the meteoritic material, they found it to be 93 per cent iron and 6.4 per cent nickel; but, more interestingly, mixed with these relatively common metals were traces of precious platinum and iridium which alone might have made the lode worth mining, had the deposit been less deeply buried.

Although the American group failed in its attempts to mine Meteor Crater, news of the operation obviously reached Russia; and the economically reeling Soviet government may have seen its meteorites as a source of quick profit. One of the members of a later expedition to the Siberian site, when the blast was still widely considered to be the result of a meteorite, was quoted as estimating enthusiastically that they might discover a metal mass worth a fortune in the Tunguska region.

The man placed in charge of the Soviet Academy's first special meteorite expedition was a remarkable thirty-eight-year-old scientist named Leonid A. Kulik, who was then doing research at the Mineralogical Museum at Petrograd (formerly St. Petersburg, renamed in 1914). Born in 1883 in Tartu in Estonia, he had studied at the St. Petersburg Forestry Institute and in the physics and mathematics department of Kazan University. In 1904 Kulik had served in the Russo-Japanese war, and in 1910 he had been arrested and tried for revolutionary activities. After being imprisoned for a period, he remained under police surveillance until 1912.

While working in the Urals as a forestry officer, Kulik met his scientific mentor V. I. Vernadsky, leader of an expedition searching for mineral deposits, and subsequently became extremely interested in mineralogy. Vernadsky described him as a "lover of minerals and nature" who was "constantly taking pictures." He predicted that Kulik would follow in the footsteps of other great scientific researchers. E. L. Krinov, a noted Soviet scientist and meteorite authority, called Kulik "a vibrant, cultured man around whom young people flocked" and an outspoken individual who was not afraid to voice his opinions when he was convinced he was right. Vernadsky arranged to have Kulik transferred from the Forestry Department to his own expedition; and eventually Kulik and his wife, Lydia Ivanova, went to work for the Mineralogical Museum of the Academy of Sciences in St. Petersburg.

When World War I broke out in 1914, Kulik was drafted and fought briefly in the Russian Army. He happened to be on a scientific expedition in the Urals and was trapped behind the lines when the civil war erupted. He then went to the Siberian city of Tomsk, where he taught mineralogy. After returning to Petrograd in 1920, he resumed his work at the Mineralogical Museum and devoted much of his time to the acquisition and study of meteorites, a relatively new discipline in which he rapidly established himself as a leading figure. With a single-minded intensity that had characterized all of his earlier work, he studied the literature available about meteorites and attempted to add to the national collection housed at the museum. It was a specialization that soon was to set him, almost by accident, on the track of the Siberian explosion; and yet, at the same time, Kulik's concern with meteorites was to mislead him by encouraging a false conclusion about the nature of the blast.

A "meteorite" in its strictest modern definition refers to any solid, natural extraterrestrial object—usually stone or iron—that strikes the earth. Until a few hundred years ago the word "meteor," from the Greek term meaning "something raised up," was the root used in words describing all phenomena associated with the air, such as lightning, clouds, snow, or rain. The present science of meteorology, or weather conditions, still retains this connotation. During the Renaissance, for instance, man's concept of meteors reflected his picture of himself as existing on a planet located at the center of a fixed universe, in accordance with Ptolemaic cosmology. In the view of Elizabethan scholars and poets, as S. K. Heninger, Jr., explains, "Meteors were imperfect mixtures of the four Elements. They were confined to the region of the Air in the sublunary universe, since weather conditions could not conceivably transpire in the immutable region beyond the Moon." In *Richard II*, William Shakespeare made use of the popular belief that the appearance of a meteor was a sign of a disturbance in the elements, a portent of disorder and evil:

The bay-trees in our country are all wither'd
And meteors fright the fixed stars of heaven;
The pale-faced moon looks bloody on the earth
And lean-look'd prophets whisper fearful change.

Between the middle of the sixteenth century and the middle of the seventeenth, the earth-centered Ptolemaic cosmology had received a death sentence from the startling findings of astronomers like Copernicus, Galileo, and Kepler; and consequently men began to look with fresh eyes at the sky. Yet there was still much confusion and ignorance about cosmic phenomena such as meteorites. Aristotle had believed they were fiery exhalations of the atmosphere; later astronomers thought they might somehow be connected with lightning. In the eighteenth century the notion that meteorites were literally "stones from space" was greeted with skepticism, despite the evidence of numerous ancient records as well as current observations to support this fact. This extraterrestrial theory was at first totally dismissed by the prestigious Academie Française, and Thomas Jefferson scoffed at a report stating that stones had fallen from the sky in 1807 onto the soil of Connecticut.

When Leonid Kulik began his study of meteorites a century later, it was clear that we live on what Harvey H. Nininger, of the Colorado Museum of Natural History, was later to call a "stone-pelted planet," bombarded daily by millions of pieces of cosmic debris from asteroids or comets; most of these fragments are so small that they do not survive the superheated plunge through our upper atmosphere and evaporate as bright "shooting stars" or "meteor showers." The identification of the gigantic Meteor Crater in Arizona by geologist Daniel M. Barringer provided concrete proof that in our past the earth had been hit by immense meteorites. Other impact sites have been discovered around the world, including the two-mile-wide New Quebec Crater in Canada and the Vredefort Dome in Transvaal, South Africa, with a width of 26 miles. The largest solid meteoritic mass, approximately 60 tons of iron, was discovered in 1920 in Grootfontein, South-West Africa.

One year later, while preparing for his expedition to locate meteorites which had fallen in the Soviet Union, Kulik received a description of a strange event that aroused his curiosity. Another investigator passed along to Kulik a page from an old St. Petersburg calendar, containing on the back a reprinted Siberian newspaper account of the fall of a meteorite. Kulik had never before heard of this fall, and he read the story with great interest.

> About 8 A.M., in the middle of June 1908 . . . a huge meteorite is said to have fallen in Tomsk several sagenes [a sagene is 7 feet] from the railway line near Filimonovo junction and less than 11 versts [a verst is two thirds of a mile] from Kansk. Its fall was accompanied by a frightful roar and a deafening crash, which was heard more than forty versts away. The passengers of a train approaching the junction at the time were struck

> by the unusual noise. The driver stopped the train and the passengers poured out to examine the fallen object, but they were unable to study the meteorite closely because it was red hot. Later, when it had cooled, various men from the junction and engineers from the railway examined it, and probably dug round it. According to these people, the meteorite was almost entirely buried in the ground, and only the top of it protruded. It was a stone block, whitish in color, and as much as 6 cubic sagenes in size.

Most of this odd report turned out to be sheer fantasy; only the detail about the train stopping near Kansk was accurate. Yet for Kulik it marked the beginning of an obsession that was to last for the rest of his life. Believing that he might have stumbled on to the discovery of a large meteorite not known to scientists, he began searching through other Siberian newspapers for further stories of the fall. He soon found buried in these papers numerous accounts of a phenomenal event that had occurred in 1908, though the details were often confusing and ambiguous.

A newspaper published in Irkutsk, 550 miles from the explosion, related that on a morning in June of 1908, in a village north of Kirensk, peasants saw "a body shining very brightly (too bright for the naked eye) with a bluish-white light. It moved vertically downwards for about ten minutes. The body was in the form of 'a pipe,' i.e., cylindrical." The paper further stated that after the bright object fell, "a huge cloud of black smoke was formed" and a crash as if from "gunfire" was heard. "All the buildings shook," the report went on, "and at the same time a forked tongue of flame broke through the cloud." The villagers had dashed out into the streets in absolute panic; some wept in terror, convinced that this must be the end of the world.

Kulik must have been both elated and somewhat puzzled. Meteorites were usually observed at night, not in the early morning, and the "pipe" shape did not sound like a normal meteoritic object. The cloud of black smoke and the flame were also baffling, unless the fall had set the taiga on fire. But this was unlikely at the height of summer when, as every Russian knew, much of the area was an impassable swamp.

Becoming increasingly fascinated and perplexed by each new revelation, Kulik pored through the musty newspapers, now more than twelve years old. An extraordinary story was unfolding in bits and pieces. An enormous "fiery object" had been seen over villages and towns throughout the Yenisei River province; some described it as moving almost horizontally from the south, and almost everybody had felt strong earth tremors and heard loud explosions.

In early July of 1908, for instance, a Tomsk reporter dispatched to Kansk to check out the rumors had stated, "The noise was considerable, but no stones fell. All the details of the fall of a meteorite should be ascribed to the overactive imagination of impressionable people." Yet the reporter qualified his statement by adding, "There is no doubt that a meteorite fell, probably some distance away, but its huge mass and so on are very doubtful." A week later the Tomsk newspaper, still dubious about the meteorite story, suggested that the event near Kansk had been an earthquake, followed by "a subterranean crash and roar as from distant firing. Doors, windows, and the lamps before icons were all shaken. Five to seven minutes later a second crash followed, louder than the first, accompanied by a similar roar and followed after a brief interval by yet another crash."

If it had been a meteorite, Kulik was sure that it must have been gigantic—greater than any that had ever fallen before in Russia and perhaps in the world, in order to have caused tremors like an earthquake. But where exactly had it fallen? According to the fragmentary newspaper reports, the phenomena of the fall had been observed over an area extending more than 500 miles. He would need more concrete information before he could pinpoint the exact site of the alleged fall.

One of the most specific and dramatic accounts came

from a Krasnoyarsk paper of 1908, which stated that in several villages along the Angara River, in the heart of the taiga, people saw "a heavenly body of fiery appearance cut across the sky from south to north. . . . when the flying object touched the horizon a huge flame shot up that cut the sky in two. . . . The glow was so strong that it was reflected in rooms whose windows faced north. . . . On the island opposite the village horses began to whinny and cows to low and run wildly about. One had the impression that the earth was just about to gape open and everything would be swallowed up in the abyss."

The unexpectedness and magnitude of the event seemed to have created a sense of superstitious dread among frightened villagers throughout central Siberia; after days of absorbing the newspaper reports, which tended to corroborate one another on many basic details, Kulik was not surprised that some Siberians had thought "the abyss" had opened that morning in 1908. What was surprising to him was the fact that this remarkable event had not yet been studied by any scientists—a mistake he intended to remedy on his upcoming expedition.

In a preliminary report outlining what he had learned thus far, Kulik listed the event as the "Filimonovo meteorite," since according to the first story he had read on the back of the calendar page, a train engineer had stopped at the Filimonovo junction after seeing the meteorite. At the station town of Kansk he expected to find witnesses who might clarify what had really happened and help him begin the difficult task of locating the fall point.

For the 1921 expedition, which his friend Vernadsky had persuaded the Academy of Sciences to finance, Kulik was given a railway carriage on the Trans-Siberian Express. He and his researchers left Petrograd in September, traveled across the Urals into Siberia, then made stops in Omsk, Tomsk, Krasnoyarsk, and finally Kansk. It was a long and tedious journey on a railway line still not fully recovered from the effects of war and suffering from a shortage of engineers; as few relief drivers were available then, the "express" would sometimes grind to a stop while the exhausted engineer caught up on his sleep. Kulik was reminded of his earlier trips to the Ural Mountains and to Tomsk which, like most of the stops along the railroad line, was hardly more than a small country settlement with none of the heavy industry that the Soviets were to construct in later decades.

At Kansk, though he quickly found that he was nowhere near the site of the 1908 explosion, Kulik was able to check out the story about the event as experienced along the Trans-Siberian Railway; the station agent had felt a "strong vibration in the air" and heard a loud "rumbling sound," and a locomotive engineer had become so frightened by the ground tremors and noise that he had halted his train, fearing it might be derailed. Investigators had eventually arrived from Tomsk and Irkutsk but had found no sign of a meteorite. As a result of a questionnaire he circulated in Kansk and the surrounding districts, Kulik collected a large number of remarkably vivid personal recollections of the incredible luminous phenomena and the destruction that occurred early in the morning of June 30, 1908. From these eyewitness testimonies, Kulik decided that the fiery object must have impacted farther north, near the basin of the Stony Tunguska River.

Despite the fact that many of the details he acquired conflicted with the traditional signs of such a fall, he was convinced that the object had been a meteorite. Seeking the elusive proof of this became in time a life's work. But for now, he and his team had to return to Petrograd, and the search would have to wait six years.

Four

THE TUNGUSKA EXPEDITION

Almost immediately after arriving back in Petrograd, Kulik began thinking about the next and more important expedition into Siberia. Following his report to the Soviet Academy of Sciences on the inconclusive 1921 journey and during the next six years, he received further data from other investigators and additional eyewitness stories that made the explosion seem even more potent than he had imagined. These reports confirmed his belief that its epicenter, or fall point, lay north in the region of the Stony Tunguska; and he soon became convinced that a thorough survey of this area, preferably in the early spring when the weather would be at least tolerable, would uncover the true nature of the strange detonation and enable him to sort out fact from fiction in the numerous rumors circulating about the event.

Several other scientists who happened to be working in the Tunguska region gathered intriguing and sometimes frightening tales from the local inhabitants, the Tungus.

S. V. Obruchev, a geologist conducting research along the Stony Tunguska River in the summer of 1924, encountered such superstitious awe among the natives about the blast, which he presumed had been caused by the impact of a large meteorite, that he wrote, "In the eyes of the Tungusi people, the meteorite is apparently sacred, and they carefully conceal the place where it fell." As Kulik was later to discover on his second Siberian journey, many Tungus were afraid to talk about the explosion and some completely denied its existence. Others reluctantly admitted to Obruchev only that a huge area of "flattened forest" could be found by traveling three or four days northeast of Vanavara to a wild and almost inaccessible part of the country near the Chambé and Khushmo rivers. Another local report sent to Kulik stated that, according to the Tungus, at least a thousand reindeer had been killed and several of their nomadic villages had vanished during the explosion. "A violent wind" had leveled the taiga, said others, and "water broke from the earth."

One of the most striking accounts of the effect of the explosion came from Ilya Potapovich, a Tungus who later became the chief guide for the 1927 expedition. Ilya Potapovich's harrowing story about his brother's experiences was recorded in 1923 and sent to Kulik by a geologist named Sobolev, who was working near the area:

> Fifteen years ago his [Ilya Potapovich's] brother, who was a Tungus and could speak little Russian, lived on the River Chambé. One day a terrible explosion occurred, the force of which was so great that the forest was flattened for many versts along both banks of the River Chambé. His brother's hut was flattened to the ground, its roof was carried away by the wind, and most of his reindeer fled in fright. The noise deafened his brother and the shock caused him to suffer a long illness. In the flattened forest at one spot a pit was formed from which a stream flowed into the River Chambé. The Tunguska road had previously crossed his place, but it was now abandoned because it was blocked, impassable, and moreover the place aroused terror

among the Tungusi people. From the Podkamennaya [Stony] Tunguska River to his place and back was a three-day journey by reindeer. As Ilya Potapovich told this story, he kept turning to his brother who had endured all this. His brother grew animated, related something energetically in Tungusk language to Kartashov, striking the poles of his tent and the roof, and gesticulating in an attempt to show how his tent had been carried away.

According to the Tungus brother's widow, Akulina, who was questioned in 1926 by ethnographer I .M. Suslov, the entire family in the tent was thrown into the air and several knocked out by the explosion. The tent was approximately 25 miles southeast of the blast site. When Akulina and her husband woke up, Suslov reported, they saw "the forest blazing around them with many fallen trees. There was also a great noise." Suslov spoke to an elderly Tungus who had been sharing the tent with the family and recorded this story:

> Vasily had been sleeping at the moment when the tent was torn away and had been thrown to the side by a powerful jolt. He had not lost consciousness. He said that he heard an unbelievably loud and continuous thunder; the ground shook, burning trees fell, and all around there was smoke and haze. Soon the thunder stopped, the wind ceased, but the forest continued to burn. All three of the Tungus went out to search for the reindeer which had run away during the catastrophe. But they were unable to find many of the reindeer.

As the eyewitness tales continued to pour in, it became apparent why the Tungus regarded the 1908 catastrophe as a divine punishment, the inexplicable wrath of a vengeful god. Suslov, who was studying the culture of the people of northern and central Siberia and had established a rapport with the Tungus, constantly encountered tribesmen with horrifying tales of destruction. At Strelka, a tiny trading post on the Chunya River, he met the Tungus

Podyga's children who had been living in a tent by the Avarkita River at the time of the explosion. "A terrible storm, so great that it was difficult to stand upright in it, blew down the trees near their hut," Suslov was told. "In the distant north, a large cloud formed which they thought was smoke."

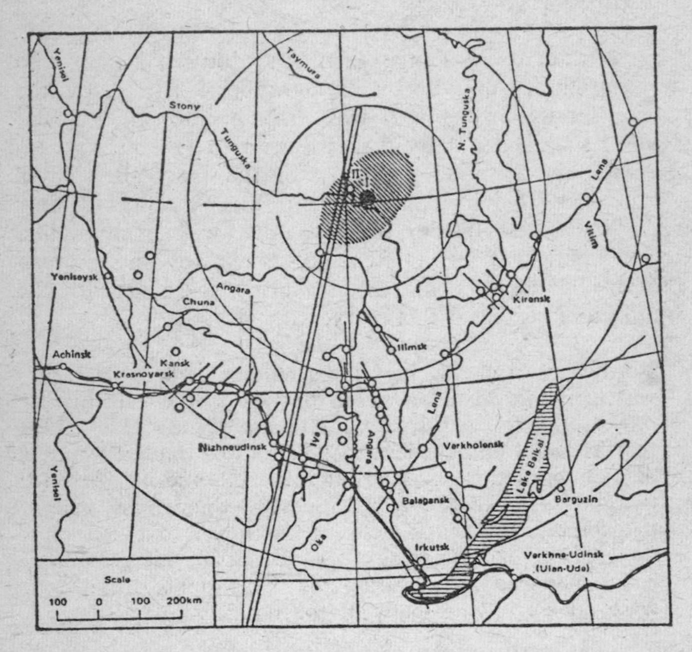

A. V. Voznesensky's chart, published in the mid-1920s, showing his estimate of the extent of the blast phenomena observed throughout central Siberia. The shaded area marks the presumed location of the forest destruction; lines through towns indicate the direction from which eyewitnesses saw either the "fireball" or the blast's effects. On the basis of these reports and seismic data, Voznesensky calculated a possible trajectory indicated by a double line running from the south-west. (Courtesy of Pergamon Press)

At Strelka Suslov talked with a group of about sixty Tungus who agreed not only that the 1908 catastrophe had "crushed" the taiga, killing their animals and injuring some of their people, but also that the blast had "brought with it a disease for the reindeer, specifically scabs, that had never appeared before the fire came."

The 1908 fireball's direction of flight and the probable location of the blast had been estimated in the mid-1920s by A. V. Voznesensky, former head of the Irkutsk Observatory. Using some of the recent information acquired by Kulik and Obruchev, as well as earlier seismic data from Iskutsk and other Russian stations and observations of acoustical phenomena throughout central Siberia, he attempted to trace the path of the body and determine its impact point. He found that the effects of the explosion had been seen and heard by people over an incredibly immense geographical area, one larger than France and Germany combined. The "fiery object" racing through the cloudless sky had been observed by thousands from the southern border of Siberia to the Tunguska region, while the noise of the explosion, the heavy claps, and rumblings "like thunder" were audible for a radius of 500 miles. From these reports and the seismic data, he was able to gauge the time of the blast at about 7:17 A.M. on June 30, 1908. The place of the fall, he determined, was in the territory north of Vanavara.

Voznesensky speculated that the explosion was caused not by a single meteorite but rather by a group which "were flying in the same direction and gradually breaking up." The tremendous aerial waves were caused as the "fragmentations of the meteorite" expended their energy over central Siberia, and the blast and ground vibrations were caused by "a very considerable mass that fell on to the ground." Convinced that a future expedition would discover in the Tunguska region a gigantic impact hole similar to Arizona's Meteor Crater, Voznesensky concluded that

> . . . it is highly probable that the future investigator of the spot where the Khatanga [Stony Tunguska] meteorite fell will find something very similar to the meteorite crater of Arizona; i.e., from 2 to 3 kilometers around he will find a mass of fragments that were separated from the main nucleus before it fell and during its fall. The Indians of Arizona still preserve the legend that their ancestors saw a fiery chariot fall from the sky and penetrate the ground at the spot where the crater is; the present-day Tungusi people have a similar legend about a new fiery stone. This stone they stubbornly refused to show to the interested Russians who were investigating the matter in 1908. However that may be, the search for and investigation of the Khatanga meteorite could prove a very profitable subject of study, particularly if this meteorite turned out to belong to the iron class.

The detailed eyewitness testimony gathered by Obruchev and Suslov and the calculations of Voznesensky were helpful to Kulik in persuading the Soviet Academy, despite the fact that many of its members remained unconvinced that the earlier evidence indicated a meteorite, to authorize the first expedition to the Stony Tunguska River basin. Thus, in February 1927, Kulik departed from Leningrad (as Petrograd was renamed in 1924) with a research assistant and traveled by Trans-Siberian Express to Kansk and then farther east to the remote station of Taishet. In each of these small outposts he met more people who agreed on the northward direction of the "ball of fire" and who had heard the "prolonged thunder" of its explosion. In Kansk, 375 miles southwest of the Stony Tunguska area, persons in the streets had "felt a subterranean rumble. Inside, hanging objects were noticed to rock, china was smashed, and in one house, the inside wooden shutters rattled."

Other reports from the Kansk district told of boatmen being thrown into the river and horses knocked down by the force of the thunder and the ground tremors. At Taishet, about a hundred miles farther along the railway line, buildings and telegraph poles shook, doors banged in houses, and objects fell to the floor.

In March, after gathering supplies and equipment, Kulik and his assistant began their journey northward from Taishet toward the presumed fall point. Even this far into spring, central Siberia was one of the least comfortable places on earth, but Kulik had little choice about the schedule. If he had come earlier, snow would have made the region impassable. By mid-summer, much of the taiga turned to marsh where clouds of mosquitos tormented men and animals. But in March, with the ground still firm under a light cover of snow, it was possible to move through the area, although occasional storms reducing temperatures to minus 40 degrees Fahrenheit could make any expedition an endurance test.

Kulik and his assistant traveled by horse-drawn sled along the Angara River to Keshma, a small village where they purchased more food and supplies. The country soon became more rugged, split with creeks, gullies, and steep hillsides. The high latitude confused their compasses. Maps were inaccurate, where they existed at all. Finally, near the end of March, they arrived at Vanavara on the Stony Tunguska River. The last stop before proceeding into the vast taiga, the tiny settlement consisted only of a few houses, trading stores, and muddy streets. Kulik hired Ilya Potapovich, the Tungus, as his guide and with his assistance began questioning local residents about the explosion.

From several inhabitants of Vanavara Kulik obtained remarkable stories of the blast, particularly of the searing heat and shock waves. On the morning of June 30, 1908, S. B. Semenov, a farmer, had been sitting on the open porch of his house, looking toward the north when suddenly he saw a "great flash of light" above the northwestern horizon. He wrote of the event:

> There was so much heat that I was no longer able to remain where I was—my shirt almost burned off my back. I saw a huge fireball that covered an enormous

> part of the sky. I only had a moment to note the size of it. Afterward it became dark and at the same time I felt an explosion that threw me several feet from the porch. I lost consciousness for a few moments and when I came to I heard a noise that shook the whole house and nearly moved it off its foundation. The glass and the framing of the house shattered and in the middle of the area where the hut stands a strip of ground split apart.

At the same time a neighbor, P. P. Kosolapov, who had been working outside by the window of the house, felt his ears burned as if by a "powerful heat." He put his hands over his ears and asked Semenov whether he had seen anything. "How could one help but see it?" Semenov replied. "I felt as though I had been seized by the heat." Kosolapov went inside the house when abruptly "there was a great clap of thunder and the sod poured from the ceiling, a door flew off the Russian stove, and a piece of window glass fell into the room."

Although Kulik had never heard of such heat phenomena associated with a meteorite fall, he believed that this could be accounted for by the meteorite's size and the release of energy in its collision with the earth. He was wrong, but at this time neither he nor any other scientists had sufficient knowledge to explain this type of radiant burning.

Like earlier investigators, Kulik also found that some Tungus preferred not to discuss the event. A few were openly hostile. Gradually he learned the details of a new religion that had sprung up among some inhabitants of the taiga since the explosion, one which made the Tungus unwilling to help anyone approach the fall site. The fiery body, they claimed, was a visitation from the god Ogdy (Fire), who had cursed the area by smashing the trees and killing all the animals. No man dared approach the fall site for fear of being cursed by Ogdy's fire. Tales circulated of herds of reindeer being slaughtered to placate the god and of rumors that Ogdy, enraged at invasions of his territory, would hold back the thaw if visitors disturbed him.

These stories only made Kulik more anxious to see the blast area for himself. On the day after arriving in Vanavara, he and Ilya Potapovich tried to ride on horseback into the taiga but were unable to make any progress because of an unusually deep snowfall. Roads into the forest were nonexistent. They returned to Vanavara, where Kulik continued interviewing the local people and preparing for the trek into the wilderness. Then, on April 8, Kulik, his assistant, and Ilya Potapovich set out with pack horses along the winding path of the Stony Tunguska River. By the time they reached the hut of Okhchen, a friendly Tungus, on the Chambé River, all were exhausted, suffering from scurvy brought on by the lack of proper food and from infections caused by pestilential marsh over which they had to move. Kulik and his colleague had never endured such hardship, but the belief that they were almost at the end of their journey gave them the energy to keep going.

From the Tungus they learned that they were close to the beginning of the devastated forest, the edge of the blast area. After resting for one night, they loaded their equipment onto reindeer and set off along the bank of the Chambé, then left the river and proceeded directly north. Within two days, they crossed the Makirta River which flows into the Chambé. It was April 13. At the edge of the Makirta the small expeditionary party stared at an incredible sight: the first signs of the enormous force of the 1908 explosion.

Five

FALL POINT

"The results of even a cursory examination exceeded all the tales of the eyewitnesses and my wildest expectations," Kulik later wrote of his discoveries in the Tunguska region. Standing on the sloping bank of the Makirta River, the first scientist to view the actual physical effects of the explosion, he gazed in stunned amazement at the sight. Nothing he had heard could have prepared him for the immensity of the devastation.

For as far as Kulik could see, both upstream and down, the upper part of the riverbank was littered with the trunks of fallen birches and pines, apparently smashed down by the shock wave of the blast. The lower part of the bank where the forest still stood was cluttered with undergrowth that included more trunks and the decayed limbs of trees. Against the blanket of snow, the uprooted trees and broken branches were outlined in stark relief. Small knolls along the bank, Kulik later wrote, stood out "picturesquely against the sky and taiga, their almost

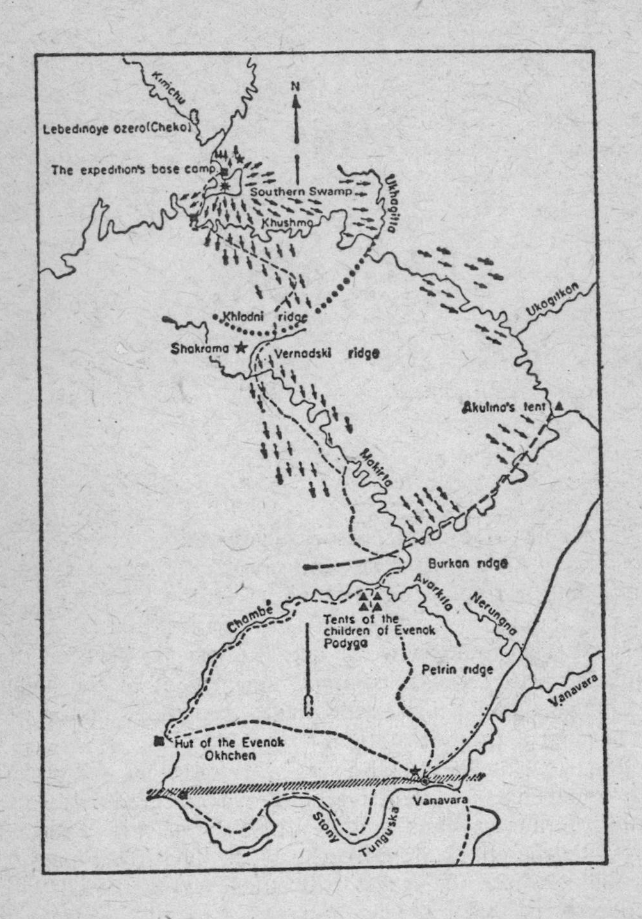

The route of Kulik's first expedition from Vanavara to the Southern Swamp, with dots showing the limits of the scorched area and arrows indicating the direction of the fallen trees. (Courtesy of Pergamon Press)

treeless snow-capped tops stripped bare by the meteorite whirlwind of 1908."

The weary party marched northward through ever greater devastation. The number of uprooted trees increased, their tops always pointing toward the south, the direction in which they had been heaved by the blast. In some places entire stands of giant larches several hundred years old had been knocked down; the ground was littered with fallen, dead trunks, their roots torn from the earth, their foliage stripped away. Often the group had to hack its way through the entangled limbs.

Eventually the party reached an area where the dead trees bore, in Kulik's words, "traces of a continuous burn from above." Even the broken limbs of those trees still upright were charred at the break. The fact that every broken branch showed signs of fire indicated that the burns were not those of a forest fire but the result of a sudden, instantaneous scorching—a flash of intense heat that seared and charred everything. As they progressed deeper into the taiga, the signs of heat flash increased. At first Kulik thought the scorching might have been caused by a "hot, compressed-air pocket" pushed before a meteorite. In a later report he summarized his early speculations:

> Before it actually hit the ground, the great swarm of meteors must have traversed two hundred or three hundred miles of the earth's air. Pushed ahead of it was doubtless a giant bubble of superheated atmosphere, hotter than the blast of any earthly furnace and under the enormous pressure produced by the meteor's flight. That white-hot air blast was probably responsible for the burnt spot which surrounds the place where the meteorite lies.

After several more days of travel, Kulik made his way up one of the taiga slopes. Atop the Khladni Ridge he could see the entire area for miles in all directions and, for the first time, gain an impression of the total extent of the destruction.

The view before him was astonishing. The ridges of all the lower surrounding hills were stripped bare. As far as the eye could see stretched enormous dark patches of scorched, flattened forest—trees uprooted and laid down at the same angle, their tops facing south or southeast, the dead trunks with their exposed roots pointing directly north. Only in some of the deeper protected valleys to the east and west had the forest survived.

Looking due north from the ridge for 6 or 7 miles to the horizon, Kulik could see little remaining of the great taiga.

> I still cannot sort out my chaotic impressions of this excursion. Above all, I cannot really take in the whole majestic picture of this unique meteorite fall. A very hilly, almost mountainous, region stretches away tens of versts towards the northern horizon. In the north the distant hills along the River Khushmo are covered with a white shroud of snow half a metre thick. From our observation point no sign of forest can be seen, for everything has been devastated and burned, and around the edge of this dead area the young twenty-year-old forest growth has moved forward furiously, seeking sunshine and life. One has an uncanny feeling when one sees 20- to 30-inch thick giant trees snapped across like twigs, and their tops hurled many metres away to the south. This belt of verdure surrounds the burnt area for some tens of versts to the south, southeast, and southwest of the observation point. Nearer the periphery, the verdure gradually merges into the ordinary taiga, the extent of the windbreak soon decreases and falls to zero. Only in places on summits and hilltops with a more-or-less normal forest wall facing the air current does an area of prostrated mature forest stand out now as a white patch; and beyond is the taiga, the endless, mighty taiga, for which terrestrial fires and winds hold no terrors, for they leave no greater scars than scratches on the hands and face of one of its people.

Eager to push on immediately toward the center of the

blast area, which he assumed lay beyond the distant northern ridges, Kulik urged his guides, Ilya Potapovich and Okhchen, to lead him straight through the burned-out taiga. But the superstitious Tungus, obviously fearful of the punishment which would follow this invasion of accursed territory, refused to go on. Kulik could not induce them to go further north, and he found himself forced, this close to his goal, to retrace his steps to Vanavara and recruit another group of guides.

On April 30, after finding new helpers, Kulik and his assistant left the trading post and traveled by sled for three days back to the Chambé River. Now painfully aware of the difficulty of marching through the ruined forest, he and the Tungus built rafts with which they worked their way up the swollen, rapids-ridden Chambé and Khushmo rivers toward the area of destruction. Then, loading themselves up like pack animals, the party headed north on the last leg of what had become an epic journey. Not until May 20 did they finally arrive again at the area of the devastation. This time Kulik was determined not to return until he had found the fall point.

In another week of arduous travel, during which they often had to cut their way through the entangled mass of fallen, charred trees, Kulik set up camp in a valley near the mouth of the Churgima River. He was now far beyond the Khladni Ridge from which he had first surveyed the region and within reach, he estimated, of the point where the huge meteorite had crashed to earth. To the north, beyond the ridges and hills stripped of trees lay, according to his guides, a large marshy basin called the Southern Swamp, which must be the site of the enormous crater he had come to expect. Using this camp as a base, Kulik made daily trips into the shattered forest, until he had circled the entire area.

It was now early June, more than three months since the start of Kulik's inadequately equipped expedition—one which would soon become among the most famous and controversial in the annals of modern scientific investigation. In Leningrad, some of his scientific colleagues, having heard nothing from him since the beginning of his trip, had begun to worry about his safety, fearing that he might have become lost or even died in the uncharted Siberian wilderness; a relief expedition to search for him was considered. Other scientists had not taken his absence very seriously. Based in their opinion on unreliable rumors and tales heard from Tunguska peasants rather than on verifiable data, the trip, they assumed, was taking longer than anticipated simply because Kulik had been unable to find any real evidence of the phenomena for which he searched.

Completely absorbed in his mission, Kulik had no idea that his colleagues back in civilization were wondering whether he might be dead. He had already found and photographed enough real proof of a cataclysmic event to amaze and puzzle the Soviet scientific establishment for decades. His main difficulty as he roamed through the shattered forest around the Southern Swamp, which he now mentally dubbed the Great Cauldron, was in deciding just what the evidence indicated.

> At first I went towards the west, covering several kilometres over the bare hill crests; the treetops of the windbreak already lay facing west. I went round the whole cauldron in a great circle to the south, and the windbreak, as though bewitched, turned its treetops also to the south. I returned to camp and again set out over the bare hills to the east, and the windbreak turned its treetops in that direction. I summoned all my strength and came out again to the south, almost to the Khushmo: the tops of the windbreak also turned towards the south.

After having made his way around the entire rim of the frozen swamp, about which trees lay flattened radially like a vast fan, Kulik knew he had reached his goal. "There could be no doubt," he wrote, "I had circled the center of the fall!" He stood on the edge of the region that, according to Tungus lore, had been cursed by the fire god. Yet it was an area that did not agree with any he had ever seen, and of the vast meteorite crater he expected there was no sign at all.

His notes betray his confusion and a desperate attempt to cling to the meteorite hypothesis which had brought him to the site. A single observer on the ground, clambering over the tangled trunks of huge trees and cutting his way through dense undergrowth, he saw the fall point only in a series of phantasmagoric vistas, flashes which he recorded accurately but which would not be put into context until expeditions decades later, armed with helicopters and sophisticated recording equipment, created a detailed picture of the area.

In places, the taiga was leveled like that of the area surrounding the fall point. "The taiga," Kulik recorded, "both in the cauldron and outside it, has been practically destroyed by being completely flattened. It lies in roughly parallel rows of bare trunks without twigs and tops, their upper ends turned away from the center of the fall."

But in another part of the woods, Kulik was astonished to see upright trees gradually begin to appear among the tumbled debris and the crushed forest with its dead trunks and exposed roots give way to one infinitely more eerie. Near what he was convinced must be the epicenter of the fall, he found himself back once again in a standing forest. Stripped of bark and limbs, as dead as those which lay in thousands on the ground for miles around, a zone of trees in the cauldron was miraculously preserved.

Kulik could only speculate about some freak effect of the shock wave which had left an area of trees protected from the blast. Sketching the phenomenon for a journalist a few years later, he guessed it to be "some kind of node or region of rest due to the interference of air waves," but the dead, upright forest remained one of the most inexplicable discoveries of the expedition.

Beyond the zone of standing trees, which seemed to Kulik to exist in a huge ring around the actual impact point, he found what they had traveled thousands of miles to see—the peat bog in which, he guessed, the vast meteoritic mass had landed. It resembled a scene of utter devastation. "The peat marshes of the region are deformed," Kulik wrote, "and the whole place bears evidence of an immense catastrophe." Several miles of swamp appeared to have been blasted and tortured into a landscape like the fantasy of a surrealist painter. "The solid ground," read one of Kulik's reports, "heaved outward from the spot in giant waves, like waves in water," and he supposed it "actually must have splashed outward in every direction." Beneath a thin layer of new moss and vegetation, the marshy region showed marks of "uniformly continuous scorching, unlike the traces of an ordinary conflagration." In the northwest and northeast sections of the burnt area, Kulik found dozens of "peculiar flat holes" ranging from several yards to dozens of yards in diameter and several yards in depth.

The single, huge impact crater that Voznesensky had predicted did not exist. The shallow holes that pitted the area, Kulik assumed, must mark the fall point of individual fragments of a meteorite. Most of the holes were overgrown with moss and a few were filled with frozen water, but nature's partial disguise did not obscure the fact that they were unlike any known meteorite impacts. Kulik first theorized that "with a fiery stream of hot gases and cold, solid bodies, the meteorite had struck the cauldron, with its hills, tundra, and swamp and, as a stream of water striking a flat surface splashes spray in all directions, the stream of hot gases with the swarm of bodies penetrated the earth and both directly and with explosive recoil wrought all this mighty havoc."

This initial dramatic explanation of the waves or folds in the earth and the craterlike holes scattered about the peat bogs of the Southern Swamp was to be questioned by his fellow scientists and put in serious doubt by the realities of other recorded meteorite falls. Kulik had no satisfactory explanation for the upright trees in the "telegraph forest," later to be revealed as an essential physical clue pointing to the Tunguska explosion's true nature. Soon he would be forced to re-examine his early theories, though he never completely surrendered his notion of the meteoric origin of the blast. On subsequent expeditions he continued to probe beneath the taiga's permafrost for evidence of a meteorite, but always in vain.

The tiny group had brought no equipment to bore into the pits nor to take precise measurements of the region. They could do little more than make a hasty survey and draw a rough map of the locale. Because of the dangerous summer thaws, remaining longer in the wilderness would have been hazardous. Kulik turned his thoughts to returning home and presenting the evidence, contradictory and confusing as it was, to his fellow scientists.

"We had food left for only three or four days," he wrote, "our road was a long one and our one thought was to get back safely. It was flight in the full sense of the word. We were already living on the remains of our food supplies (our expectation of obtaining game had not been realized), cutting down our rations as much as possible, and shaking out the flour bags. Three or four times we shot duck and once or twice caught a fish, but there was little to be found except puchki [a Siberian plant with edible stems]. For nine days we traveled day and night down the Khushmo and the Chambé to the Podkamennaya Tunguska. Our diet consisted almost entirely of puchki, and we speculatively estimated the weight of our last reserve—the horse."

In late June, preserving "some remains of courage," the exhausted group finally reached the Stony Tunguska River near Vanavara just as a light rain was falling.

Six

THE RIDDLE

To the surprise of his debunkers, Kulik returned from the grueling ordeal of the 1927 expedition in triumph—or at least what seemed like triumph at the time. The solid evidence he had acquired impressed the Soviet Academy of Sciences and quashed the skepticism and indifference with which many scientists had previously greeted stories of the strange superblast. A catastrophic explosion rumored to have occurred almost two decades ago in the remote forests of central Siberia became, because of Kulik's first-hand documentation, an undeniable and even somewhat overwhelming reality that quickly aroused the fascination of experts not only in Russia but around the globe.

As his accomplishments were publicized, Kulik's reputation gradually changed from that of a relatively obscure meteorite researcher to that of a pioneering figure of growing international prestige. His report delivered in December of 1927 to the Presidium of the Academy of Sciences reflected his own picture of his place in history and his stubborn belief in the importance of financing immediate further investigation of the remarkable event:

> For seven years I have been holding the view that since this fall occurred on the territory of the Soviet Union, we are in duty bound to study it. If the matter was delayed until last year on the pretext that it was all pure fancy, this objection has now been swept away, since the positive results of my expedition are irrefutable. Their unique scientific significance, like the significance of the Tunguska fall itself, will be fully appreciated only in history and it is necessary to record all the remaining traces of this fall for posterity.

In addition to extensive coverage in Russian and European newspapers, articles appeared in the New York *Times* and the London *Times* featuring the Tunguska expedition and quoting Kulik's descriptions of the "blasted" and "bombarded" impact site. The mysterious explosion also received considerable attention in popular astronomy and science magazines as well as serious scientific journals. As the event became internationally known, additional records of huge seismic shocks and air turbulence in 1908 were uncovered and sent from other countries to the Soviet Academy. The magnitude of the blast was further confirmed when it was determined from world-wide barographic records that its air waves had traveled twice around the earth.

For the next several years exploratory trips to the fall region, under Kulik's direction, became almost annual events. On the second Tunguska expedition, begun in April of 1928, Kulik was joined by zoologist V. Sytin and Sovkino cinematographer N. Strukov from Moscow. In addition to filming graphic examples of the shattered forest, Strukov shot a hazardous river crossing in which Kulik nearly drowned. The accident occurred on some narrow, rocky rapids on the Chambé River where the group was forced to unload their boats and carry all the equipment across the water; as Kulik was pushing his boat across, it overturned and knocked him down. Only the fact that his foot became entangled in a mooring line prevented him from being swept downstream. Clinging to the boat, he pulled himself to the bank and emerged safely, still wearing his spectacles.

After arriving at the Southern Swamp, the group explored a wider area around the fall point and conducted a magnetic survey, hoping to detect traces of meteorite fragments imbedded in the peat. Although some Tungus had reported finding unusual bits of shiny metal "brighter than the blade of a knife and resembling in color a silver coin," Kulik's primitive magnetic instruments detected nothing. The team tried to dig into the large circular depressions, which he was sure were craters caused by meteorite fragments, but the water and boggy soil made penetration almost impossible without boring and draining equipment.

A third and better-equipped expedition, departing in February of 1929 and remaining in the wilderness for eighteen months, also concentrated on studying the Southern Swamp, particularly the craterlike pits and the huge peat folds or ridges which Kulik believed were formed by "the enormous lateral pressure of the explosion gases which emanated from the meteorite." Despite elaborate trench digging, excavations, and boring with large hand drills to depths of 25 yards, no proof was found to support this notion. The possibility that the pits and ridges were simply natural formations of the central Siberian landscape, caused by the thawing of its permafrost beneath the peat, had occurred to other scientists, including E. L. Krinov, a colleague on the third expedition and early authority on the Tunguska blast. Krinov, who lost a toe to frostbite on this expedition, believed that Kulik might have been more successful had he not focused his attention almost exclusively on what he was convinced lay hidden beneath the marshy turf of the Southern Swamp and instead broadened his scope to the larger region of the wind-fallen taiga.

But for the next decade Kulik obstinately persisted in

his conviction that under the swamp lay "crushed masses of . . . nickeliferous iron, individual pieces of which may have a weight of one or two hundred metric tons." The original meteorite, he estimated, probably weighed, before falling into the earth's atmosphere, "as much as several thousands of metric tons." His companion on the second expedition Sytin guessed that the value of the metal might be between 100 and 200 million dollars, chiefly for the iron and platinum.

Following the 1928 trip to central Siberia, Kulik had given a lecture, accompanied by Strukov's motion picture of the Tunguska destruction, to a Moscow audience that, according to the New York *Times*, "shivered" as he outlined one of the more alarming implications of the event:

> Astronomers and geologists know that this was an exceptional circumstance. But they know also that there is no reason whatever why a similar visitation should not fall at any moment upon a more populous region.
>
> Thus, had this meteorite fallen in Central Belgium, there would have been no living creature left in the whole country; on London, none left alive in South [of] Manchester or East [of] Bristol. Had it fallen on New York, Philadelphia might have escaped with only its windows shattered, and New Haven and Boston escaped, too. But all life in the central area of the meteor's impact would have been blotted out instantaneously.

Similarly, a Moscow newspaper speculated about the harrowing possibility of a repeat of the blast:

> The fall of this meteor in the trackless Siberian forest did nothing more than provide a new legend to the local savages. But how can we know that another will not strike Moscow, London, or New York? The Siberian area was completely devastated of all life for an area of 100 square miles. In a densely populated region a similar phenomenon would be one of the most appalling catastrophes in human history.

The position of the 1908 explosion raised another chilling question. Anyone glancing at a global map could see that Leningrad is on the same latitude as the Tunguska cataclysm. Had the event happened slightly later that morning and the earth rotated only a few more hours, would this huge city have been annihilated by a direct hit from the cosmic object?

In the United States, an astronomer at the Leander McCormick Observatory at the University of Virginia, Charles P. Oliver, writing in 1928 for *Scientific American*, labeled the Siberian explosion "the most astonishing phenomenon of its kind in scientific annals" and said of Kulik's report that "one has to admit that many accounts of events in old chronicles that have been laughed at as fabrications are far less miraculous than this one, of which we seem to have undoubted confirmation."

The curator of meteorites at the Colorado Museum of Natural History, Harvey H. Nininger, urged his fellow astronomers and scientists to equip and send an American expedition to Russia in order to thoroughly evaluate the "unparalleled" Stony Tunguska event discovered but not explained by Kulik. Nininger stressed, in his 1933 book *Our Stone-Pelted Planet*, that further delay could mean the loss of invaluable physical evidence which was fading with the years and that he hoped a properly equipped group of scientific investigators would arrive in time "to secure what is yet available of the greatest message from the depths of space that has ever reached our planet."

While many researchers agreed about the unique opportunity presented by the Siberian event, Nininger was not able to gain financial support for an American expedition; and U.S. astronomers had to be content with reading about Kulik and the increasing number of other Soviets involved in the Tunguska research. Discussing with a British writer his lack of success, after three expeditions, in locating positive proof of a meteorite fall, Kulik pointed to the possibility that the pits in the Southern Swamp might be "the marks of a ricochet where the meteorites struck the earth a glancing blow and then flew off into space again, or perhaps vaporized on the spot." He also conceded that the meteorites might "have been connected with the comet Pons-Winnecke, or indeed they may have been the comet itself." But Kulik's inability to substantiate his speculations encouraged the advancement of other theories to account for the Tunguska catastrophe.

In the early 1930s two astronomers, F. J. W. Whipple, head of the Kew Observatory in London, and the Russian I. S. Astapovich, concluded independently that the celestial missile that scarred the taiga in 1908 had not been a meteorite but a gaseous comet which had left no trace of itself after impact. The absence of positive impact craters or sizable meteorite fragments at the scene of the explosion appeared at first to support this idea. The bright atmospheric phenomena following the blast could have been caused, Astapovich contended in a 1933 article, by the dust tail of the "nucleus of a small comet" rushing toward the earth and exploding with a kinetic energy force of at least "10^{21} ergs per second"—more powerful than major hurricanes, volcanic eruptions, and the most severe natural catastrophes. "It was this explosion that gave rise to the seismic and air waves," stated Astapovich, "while the high-temperature explosive wave caused the uniform scorching."

Although Astapovich came close in his estimate of the blast's tremendous release of kinetic energy—it was established in the 1960s by the Soviets to be as much as 10^{23} ergs, or the equivalent of about 30 million tons of TNT—the comet hypothesis, like the meteorite theory, turned out to be highly questionable. Originally believed for thousands of years to be burning exhalations of the air portending great misfortune, comets are now known to be cosmic bodies pursuing elliptical orbits stretching out for billions of miles around the sun. Sometimes called "dirty snowballs," comets have nuclei composed chiefly of ice, frozen methane, and ammonia; their spectacular tails, which in the case of Halley's Comet had a length of more than 90 million miles, are always pointed away from the sun, the source of the dusty tail's glow.

The brilliant object that came down over Siberia, according to almost all witnesses, followed a south to north passage, and its burning trail streamed southward. There is no record, moreover, of a comet ever colliding with our planet nor any evidence that their tails cause any unusual atmospheric or magnetic phenomena such as occurred in 1908. The passage of the earth through the tail of Halley's Comet in 1910, for example, did not precipitate any striking sunsets or provoke any magnetic disturbances; its density was less than water or even air. As they near the sun, most comets have an extremely luminous and, as the Greeks described them, "long-haired" appearance; any comet, even a small one, following an orbit that would bring it into collision with the earth would be visible to half the globe long before it touched the atmosphere; the Tunguska object, on the other hand, was sighted only in the final phase of its trajectory before it exploded.

By the end of the 1930s the riddle of the great Siberian explosion was still far from solved. The cause of the blast itself was still uncertain, despite the work of astronomers, geologists, meteorologists, seismologists, and chemists and the resources of the Soviet Academy; and many aspects of the destruction site, such as Kulik's bare "telegraph pole" forest around the Southern Swamp, remained inexplicable. Samples of soil from the fall point had been collected but not fully analyzed. Because Kulik had concentrated primarily on the central region and no expedition had yet explored the entire area of the uprooted trees, the precise borders of the destroyed taiga had not been carefully mapped out or examined. The exact shape of the explosion wave was not yet known, although Krinov had surmised from partial observations that it had an oval form.

In 1938–39 Kulik led his last expedition into the Tunguska region, principally for the purpose of taking aerial photographs of the destruction. By this time a road had been cleared through the taiga leading 40 miles from Vanavara to a camp site near the Southern Swamp, and a small air strip had been set up near the trading town. On an earlier attempt of Kulik to photograph the area from the air, the plane had crashed into the Stony Tunguska River, but Kulik and the other passengers had escaped unharmed. Although the photos finally obtained on the 1938–39 trip were not entirely satisfactory—they did not cover the whole area and were taken during the summer when the leafage partially hid the uprooted trees—they did verify the radial direction of the smashed forest and confirm the Southern Swamp as the absolute center of the explosion.

For Kulik and the other scientists who still clung to the meteorite explanation, the research of the late 1920s and 1930s led to an impasse. If the Southern Swamp was indeed the blast's epicenter, why were there no traces left of the enormous meteorite's existence? By the beginning of the 1940s Kulik's friend Krinov, in attempting to resolve this contradiction, saw the direction that future researchers must take to unravel the puzzling Siberian event:

> Thus the careful investigation of the cauldron as a whole, and of the South Swamp in particular, does not give any grounds for concluding that this cauldron is the place where the meteorite fell. But the absence, anywhere in the immediate or more distant neighborhood of the cauldron, of other areas that might attract attention as the possible places of fall, the Evenki people's designation of the cauldron as the place, the coincidence of the cauldron's co-ordinates with those of the epicentre of the seismic wave, and finally the radial forest devastation around the cauldron—all point convincingly to it as the site of the explosion. There is only one possible explanation that removes the contradiction, i.e., that the meteorite did not explode on the surface of the ground, but in the air at a certain height above the cauldron.

But the intriguing possibility that the blast may have taken place at a high altitude was not immediately investigated. All research on the Tunguska catastrophe was abruptly halted by the advent of a greater holocaust, World War II.

In December of 1938 the Soviet Academy of Sciences had passed resolutions praising "the considerable achievements of Kulik and his group during this period in elaborating a technique for finding the possible point at which the meteorite fell" and noting in particular "the great persistence and enthusiasm shown by Kulik personally over many years, in searching for the place where the Tunguska meteorite fell, persistence and enthusiasm that led to the recent concrete advance in our knowledge of the subject."

On July 5, 1941, at the beginning of the Nazi advance into Russia, Kulik joined the Moscow People's Militia, a volunteer home unit composed chiefly of older men like himself with little military training. Despite the Soviet Academy's request that, because of his achievements for the Committee on Meteorites, he be excused from service, Kulik remained in the home guard.

In October, while taking part in a battle on the front line, Kulik was wounded in the leg and captured by the advancing German Army. Imprisoned in a Nazi camp in Spas-Demensk, in the Smolensk district, the fifty-eight-year-old scientist contracted typhus and died on April 24, 1942.

The leading investigator of the Siberian explosion and a pioneer in the development of Soviet meteorite research was buried in the local town cemetery.

Seven

THE ANSWER?

Almost as if Leonid Kulik's tragic death had been a signal, speculation on the cause of the Tunguska explosion took a new and unexpected turn after World War II.

As a scientist, Kulik had been chiefly an observer and collector rather than an experimenter; he had been more inclined to accumulate and catalogue the wonders of the natural world than to inquire into their nature. With the help of a network of largely amateur collectors, who recorded new meteorite falls, Kulik had increased and meticulously catalogued the National Collection of Meteorites; and in his years of research he had done as much as one man could do to illuminate the Tunguska mystery. But the new generation of scientists had a different view and, more important, a vastly more sophisticated technology to apply to the problem. Their development accelerated by the war, communications in particular had become infinitely easier, with the result that a technologist in the Soviet Union could read of the latest theories in world science and apply them to his own field. At international conferences and on research trips, scientists of all nations observed at first hand the work of foreign colleagues and visited sites that would indirectly have a profound effect on their own work. Without such visiting between experts and pooling of research, the Tunguska riddle may have remained unsolved forever. But, in fact, the late forties brought about a revived interest in the explosion and a radical change in the theories advanced to explain the event.

Except among hard-line adherents to Kulik's theory, the word "meteorite" was seldom heard; most scientists preferred to call the Tunguska object a "cosmic body" and to describe the fall simply as an "event," a "phenomenon," or, more popularly, a "catastrophe." Every such reference suggested the new trend of speculation on what had caused the cataclysm of 1908.

One man was in no doubt over the cause of the blast. His theories, carefuly presented over the postwar years and backed up with his considerable prestige as an author and technician, were fundamentally to change the whole structure of the argument.

If one could have set a computer to choose a person to carry on Kulik's task in solving the Tunguska riddle, it is unlikely that anyone better could have been found than Aleksander Kazantsev. He had actually been born in Siberia—at Akmolinsk in 1906—and had studied at Tomsk and Omsk, cities in which speculation about the Tunguska event had a long history. Graduating from the Tomsk Technological Institute in 1930, Kazantsev, like Kulik, went to the Urals where he became head mechanic of the Beloretsky Metallurgical Plant.

Kazantsev's vivid imagination and technical skill quickly brought him to the attention of the Soviet authorities, and during the 1930s he was promoted to a post in one of Moscow's scientific research institutes where, among other tasks, he assisted in preparing Russia's exhibit for the 1939 New York World's Fair. Kazantsev joined the Army when Russia was invaded in 1941 but, too valuable to be wasted in the infantry, was soon appointed head engineer of a defense complex, where he worked on the development of new weapons. For his efforts he later received the Order of the Red Star and other honors.

Not content with his career as a technologist, Kazantsev had, long before the war, mastered chess and became an important writer on the game. In 1936 he had also exhibited another side to his multifaceted talent by entering a national competition for science-fiction film scenarios. His *Arenida* took first prize but, when it was not made into a film, he reworked it as a novel, *The Burning Island*, which was highly successful in the Soviet Union. After the war, he became a full-time author.

Like most Siberians, Kazantsev was fascinated by the bleak landscape of arctic Russia. To him, however, it represented something more than a mere frontier. He grasped the alien nature of the tundra; Mars, he felt, must look very like this frozen, wind-swept waste. Throughout his arctic travels of the mid-forties—journeys on the survey ship *Georgii Sedov* which were to serve as the basis for a series of arctic stories and fantasies—the image of northern Siberia as that part of the earth's surface most like another planet became stronger in his mind. It was to be a central concept in the evolving controversy over the Tunguska explosion.

The other impetus for new speculation on the blast came from a location less geographically remote than Mars.

In August 1945, when the American atomic bomb burst 1,800 feet above Hiroshima, the world had its first demonstration of the havoc a nuclear blast could inflict on a city. Kazantsev was among the Russian scientists who evaluated the Hiroshima data and visited the city some time after the blast. For him, the journey through its desolation had the eerie quality of a dream dimly remembered: he saw sights that were strangely familiar, phenomena that he had encountered before. Hiroshima in many respects resembled photographs he had seen of the blasted area on the Stony Tunguska where an explosion had occurred in 1908.

The Japanese explosion, made up almost entirely of flash and concussion, agreed in many ways with the damage done to the Siberian taiga in 1908 and the evidence of the eyewitnesses. At Hiroshima, only a few hundred yards from the blast center, was a group of trees, charred and stripped of their leaves but still standing upright, like those on the Stony Tunguska. Elsewhere, houses had been flattened just as the giant Siberian larches were toppled. The mushroom cloud, the black rain which fell after the blast—all were similar to what had been seen in Siberia. Every new investigation, including the detection of signs of radiation on the site, supported his theory; no meteorite or comet had caused the 1908 blast. What had exploded there was atomic.

An atom bomb in 1908? The idea at first made little sense. But to Kazantsev, with his fascination with Mars, there was only one credible explanation. An alien space ship, traveling from Mars, had chosen Siberia as a location for its landing or, more likely, plunged there out of control before exploding in the air. Why had the Martians come to earth? Kazantsev believed they came in search of water for their dying planet, and he conjectured that they may have originally been headed for Lake Baikal, the earth's largest body of fresh water.

Shrewdly, Kazantsev chose a popular magazine to publish his thesis. In America science fiction has long been accepted as a legitimate form for the presentation of scientific theories, though few scientists took advantage of its opportunities for free speculation. In Russia the tradition was stronger, and the technically oriented Soviet society of the postwar years encouraged the growth of many magazines which, like early American sf periodicals such as *Electrical Experimenter*, mixed science fiction with science fact. In 1946 Kazantsev used one of these publications, *Vokrug Sveta (Around the World)*, to put forward his theory that the devastation on the Tunguska was caused by a nuclear space ship from another world exploding high over the taiga.

In a later article, he interpreted graphically the effect of such a blast:

> The explosion wave rushed downward, and the trees directly below the point of the explosion remained standing, having lost only their crowns and branches. The wave burned the points of those breaks on the trees and hit the permafrost, splitting it. Underground waters, responding to the tremendous pressure of the blow, gushed up as those fountains seen by natives after the explosion. But where the explosion wave struck at an angle, trees were felled in a fanlike pattern.
>
> At the moment of the explosion, the temperature rose to tens of millions of degrees. Elements, even those not involved in the explosion directly, were vaporized and, in part, carried into the upper strata of the atmosphere where, continuing their radioactive disintegration, they caused that luminescent air. In part these fell to the ground as precipitation, with radioactive effects.

From another author, such a speculation would have been greeted with derision, but when presented by the distinguished Kazantsev, with his honored war record and background as a technologist, it commanded respect. Always careful to offer it only as an interesting hypothesis in the form of popular science or as science fiction, Kazantsev developed his theory for the next ten years, finally presenting it in 1958 in its most elaborate form in the story-article "A Guest from the Cosmos," which was published in *Yunyi Tekhnik (Young Technician)*, the monthly of the Communist Youth League. Later it became the central piece of his 1963 book of the same name.

"A Guest from the Cosmos" was set in a locale that, for Kazantsev, was familiar: the cabin of an arctic survey ship carrying a group of scientists into northern Siberia. This time, however, their purpose is to find a spot in the Arctic that approximates to the climate of Mars and establish whether life can exist there. One of the scientists, Krymov, claims to know with certainty that there *are*

Martians and that they have visited earth. Having been born a Tungus in the area of the Stony Tunguska, he was a boy when the great explosion of 1908 occurred. The event had far-reaching effects both on him and his family.

"My father went into the fallen taiga," Krymov explains, "and saw a huge column of water flowing out of the ground. A few days later he died in terrible pain as if he was on fire. But there was no trace of fire anywhere on his body. The old people of the tribe became terribly afraid. They forbade all of the Evenki [Tungus] people to go into the area of the fallen trees. They called it a cursed place. The shaman said that it was there that the god of fire and thunder, Ogdy, descended to the earth. All who go to that place are burned with an unseen fire."

Pressed to give his interpretation of the death, Krymov adds, "In the legend about the god Ogdy who burns with an unseen fire—what could this fire be that leaves no traces on the body? It could be nothing other than radioactivity, which begins to appear at a certain time after an atomic explosion."

To Krymov, only one explanation—that the object was a burning spaceship plunging out of control through the earth's atmosphere—could explain what they knew of the Tunguska catastrophe. "Apparently the travelers died en route from cosmic rays or from meteorite bombardment," he states. "As the uncontrolled ship approached the earth, it resembled a meteorite because it flew into the atmosphere without reducing its speed. The ship burned up from friction just as a meteorite would burn. Its outer shield was burned off, and its atomic fuel experienced conditions that made possible a chain reaction. Then an atomic explosion occurred and our cosmic guests died on the very day they were supposed to descend to earth."

Though some dismissed the idea, many members of Soviet scientific circles carefully studied Kazantsev's theories. Writing in *Znaniye-Sila (Knowledge Is Strength)* in June 1959, Professor Felix Zigel, who taught aerodynamics at the Moscow Institute of Aviation, remarked that "at the present time, like it or not, A. N. Kazantsev's hypothesis is the only realistic one insofar as it explains the absence of a meteorite crater and the explosion of a cosmic body in the air." As for Kazantsev's standing as a fiction writer, Zigel said, "It is generally known that at times—nay, often—new ideas that proved to be most valuable to science were first expressed not by scientists, but by writers of scientific fantasy." Within the Soviet Union Kazantsev's work remained highly respected and the subject of furious debate. In 1954 he was admitted to the Communist party, a considerable distinction for an author. As for his worth as a scientist, he wrote in June of 1957 an article called "Observation of Radio Signals from an Artificial Satellite and their Scientific Value" for *Radio*, the journal of the Soviet Ministry of Communications. The article foreshadowed the launching of a Soviet satellite and even revealed the radio frequency on which it would broadcast. Four months later, Sputnik 1 was launched.

As the Tunguska controversy raged on, a gradual polarization became apparent. One group, under hard-line meteorite experts like Krinov and K. P. Florensky, refused to admit that anything except a conventional meteorite could have wreaked the havoc in Siberia in 1908; even such well-publicized reversals of position as that of V. Fesenkov, who announced in October 1960 that he no longer believed in the meteorite theory, could not shake the conviction of this group. Fesenkov, A. Shternfeld, and a growing body of ingenious, often younger technologists agreed with Kazantsev that the blast had been atomic, while on the question of the power source they were, if not prepared absolutely to accept Kazantsev's idea, willing at least to keep an open mind. The postwar scientific establishment had seen enough of the atomic age and considered enough of the new theories to know that today's impossibility was tomorrow's reality. Applying the newest techniques of cosmology, atomic physics, and chemistry to the available data, they set out to seek a final explanation of the mystery of 1908.

Eight

THE FIRE CAME BY

Could the explosion of 1908 have been atomic?

Was the sudden, dazzling "flash of light" that seemed to split the morning sky over the Tunguska region the flash of a nuclear blast?

Was the painful heat experienced by witnesses 40 miles away the instantaneous thermal wave of this flash?

Could the blinding "pillar of fire" that surged upwards for miles have been an atomic fireball?

The "huge cloud of black smoke" that billowed over the entire area—was this an atomic mushroom cloud?

Was the strange disease that produced scabs on the reindeer the result of radiation burns?

These questions set Soviet scientists on a furious quest for new data that might provide the answers, as well as on a search through the mountain of evidence already accumulated by the many earlier expeditions investigating the meteorite theory. In their quest they looked for comparisons to the Alamogordo test and to the Hiroshima and Nagasaki bombs, three events of 1945 that marked the beginning of a terrifying new age.

The great explosions of antiquity belonged essentially to nature and, like most natural phenomena, almost courteously announced their coming with earthquake, storm, or subterranean fire. For the people of Hiroshima there was no such warning, nor would anyone ever again expect an announcement of impending doom. At 8:15 on the morning of August 6 the city was leveled by the blast of a uranium-235 bomb code-named "Little Boy," dropped from a height of 31,000 feet by a B-29, the *Enola Gay*, piloted by a crew to whom such a mission, despite its terrifying effects, was little more than another "milk run." American journalist David Lawrence, writing in the magazine *U. S. News & World Report,* concisely isolated the true horror of Hiroshima: "A single airplane riding high in the stratosphere, unobserved and undetected because of its great speed, can appear suddenly over London or Washington or Detroit or Pittsburgh or any city in a peaceful area and destroy human lives by the hundreds of thousands in just a few seconds." It was a new scale of destruction, excelling even the biblical promises of fire and brimstone; and among its many effects on the consciousness of man was to cause a re-evaluation of the Tunguska event. Before 1945 no frame of reference existed to explain a relatively small flying object capable of vast devastation. Hiroshima and Nagasaki provided it.

Nor was the atomic bomb merely a simpler way to make a bigger blast. The form of an atomic explosion differs materially from that created by a comparable quantity of chemical explosive. Scientists measured atomic blasts in kilotons, a unit equal to the energy output of 1 thousand tons of the explosive compound TNT, and later in megatons, 1 million tons of TNT. The Hiroshima bomb was described as the equivalent of 20 kilotons of TNT, but the comparison is misleading. A blast of TNT destroys by a sudden release of molecular energy, while the larger destructive force of an atomic explosion results from the liberation of the energy of the nuclei of atoms in a chain reaction.

A chemical explosive, such as TNT, detonates in a single massive expansion of energy, the shock wave smashing through structures, then quickly moving on. By contrast, an atomic blast builds through a chain of destruction which makes its results far more devastating. First comes the glare and heat of the thermal wave, bursting out at a temperature of 300,000° centigrade. According to physician T. P. Sears, "It moves outward as a flash with the velocity of light. It sets materials afire without contact between the material and the source of heat. It ignites wood at 1 mile and chars it beyond 2 miles. It travels faster than the shock wave; combustibles may be ignited by infrared rays and the resulting flame extinguished by the blast which immediately follows. The temperature of the ray at 4,000 feet is estimated at 1,800–2,000 C. It roughens granite, 'bubbles' tile roofs, sets gas containers on fire."

Next in an atomic blast comes the shock wave. In three seconds it travels nearly a mile, in eight seconds twice that distance. Houses within a mile are crushed by a pressure as great as 30 pounds per square inch; the average one-story brick house can be demolished by forces between 3 and 6 pounds per square inch.

A chemical explosion relies almost entirely upon the initial blast wave and concussion for its destructive effect. But in a nuclear blast, the horror is merely beginning.

The atomic bomb's primary shock wave is followed by a secondary wave as the first blast, reflected from the ground and from clouds or heavy air masses above the target, crushes down once more, distorting and flattening structures already severely damaged. This second, reinforced wave can be as much as six or eight times greater than the first primary shock.

Meanwhile, inside the atomic fireball, the upward rush of superheated air and the fury of atomic fission have created a vacuum which demands to be filled. Air is drawn into the rising globe of fire, tearing apart any buildings in its range not already collapsed, whipping debris, both inanimate and human, into the air. Set afire by the thermal wave, blazing material is sucked upwards, creating the self-feeding holocaust of a firestorm.

All these phenomena—the thermal wave, the primary and secondary shock waves, the vacuum damage, and final firestorm—left indelible marks both in the Hiroshima and the Siberian blast.

Exploded 1,800 feet in the air for maximum effect, the half-mile-wide fireball over Hiroshima expanded so as hardly to touch the ground. About one third of the output of an atomic bomb is thermal, and 25 per cent of the Hiroshima casualties suffered from flash burns. "Shadowing effects were common," read Sears's medical report. "Profile burning of the side of the face directed toward the explosion was often observed. A man writing at a window but otherwise shielded received burns of the hands. Any material whatever acted as a shield. Clothing was of great protective value beyond the 4,500 foot radius. Like sunlight, infrared is absorbed to a greater degree by dark colors than by light colors. Dark designs and polka dots were burned out of garments and tattoo burns were found on the skin beneath."

Scientists examining the ruins of Hiroshima emphasized the overwhelming effect of heat and light on the city. They found, wrote John Hersey in *Hiroshima*, "that mica, of which the melting point is 900° C., had fused on granite gravestones three hundred and eighty yards from the center . . . and that the surface of grey clay tiles of the type used in Hiroshima, whose melting point is 1,300° C., had dissolved at six hundred yards; and, after examining other significant ashes and melted bits, they concluded that the bomb's heat on the ground at the center must have been 6,000° C."

Most survivors recalled the bomb as "a noiseless flash," others as "a sheet of sun." Hersey notes that few in Hiroshima heard the bomb's noise, but all saw the vast, blinding glare and felt the wave of heat, which were followed closely by the roar of the explosion and its shock.

1. Leonid Kulik (1883–1942) discovered the impact site of the great explosion of 1908 near the Stony (Podkamennaya) Tunguska River in central Siberia and spent much of his life trying unsuccessfully to prove that it had been caused by a meteorite.

2. Ilya Potapovich (Liuchetkan), a Tungus who witnessed the frightening effects of the Siberian explosion and later served as a guide for Kulik on his first expedition in 1927 into the region of the catastrophe.

3. An expedition on the Chambé River, near the mouth of the Makirta River in the Tunguska region. In the background is the taiga, the vast forests of central Siberia, as it appears in the summer.

4. Members of an expedition crossing the Khushmo River. Mosquito nets protect their faces from fierce swarms of insects.

5. In 1927, from the Khladni Ridge, Kulik had his first long-range view of the devastated forest. In the distance the areas where the taiga has been flattened and stripped of foliage stand out as white patches of snow. Younger pines have sprouted throughout the dead forest.

6. The partially destroyed Tunguska forest seen by Kulik in 1927 as his expedition approached the fall point. Trees on the hill in the background have been leveled, their roots pointing northward toward the blast.

7. An example of the variety of devastation. Trees in a small valley have been stripped of branches but remain standing, while others on a slope have been knocked down. In the foreground is a totally uprooted tree.

8. Five miles from the blast's epicenter large trees on a hillside have been scorched and completely flattened by the heat and shock wave of the explosion. Because the blast occurred in the air, exposed hillsides suffered the greatest damage.

9. The area of peat bog known as the Southern Swamp, which Kulik called the "Great Cauldron." The explosion took place in the air, approximately two to three miles directly above this point. In the foreground is the puzzling "telegraph pole" forest, seared and stripped of branches yet still standing.

10. A closer view of the Southern Swamp, the center of the destroyed area, taken in early spring by the 1930 expedition.

11. The northwestern portion of the Southern Swamp in the summer of 1929. Small depressions found in this area were mistakenly identified by Kulik as meteorite craters; later they were found to be natural formations of the peat bog caused by the rapid melting of the Siberian permafrost.

12. Kulik set up the winter quarters of his 1929 expedition on a hillside near the Southern Swamp. New growth in the foreground already rivals in size the older trees which perished in the blast.

13. Glowing ionized air particles after the explosion created vivid atmospheric displays, photographed here, over Russia on the night of June 30–July 1, 1908.

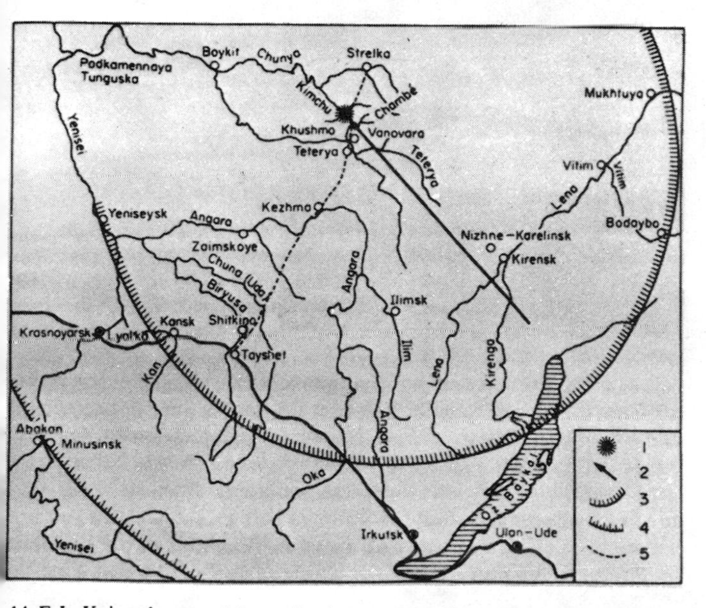

14. E.L. Krinov's map of the region between the Stony Tunguska and Lake Baikal made after his 1929–30 expedition to the destroyed area, lying at latitude 60° 55′ N., longitude 101° 57′ E. He proposed a southeast to northwest flight path for the mysterious "cosmic object" that caused the explosion. Map symbols: (1) blast site, (2) flight path, (3) extent of visual phenomena, (4) extent of region over which blast was heard, (5) 1929–30 expedition route.

15. A painting, by P.I. Medvedjev, of the Sikhote-Alin meteorite, which fell on February 12, 1947, in the mountains near Vladivostok in eastern Siberia. The largest known meteorite fall of this century, it fragmented in the air, leaving about thirty small craters impacted with numerous iron fragments. Despite its size, the Sikhote-Alin meteorite caused no radiant burning nor massive forest destruction such as was caused by the Siberian explosion of 1908.

16, 17. The famous Meteor Crater at Canyon Diablo, in the Arizona desert near Winslow, was caused by the impact of a gigantic meteorite during prehistoric times. The crater is three quarters of a mile wide and 570 feet deep. Though Kulik believed that the Siberian explosion may have resulted from a similar huge meteroite, no crater was ever found in the Tunguska region.

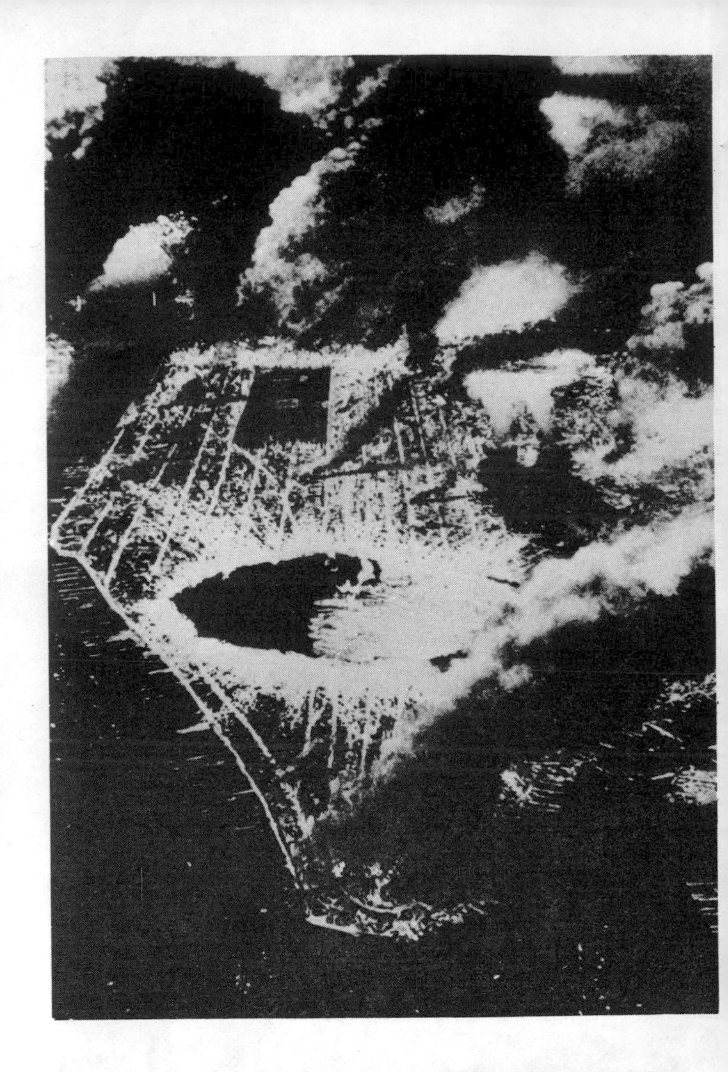

18. Artist Chesley Bonestell's impression of the effect of a giant meteorite strike on Manhattan. The meteorite is assumed to be about the same size as the one that created the Meteor Crater in Arizona. A blast with the energy yield of the Siberian explosion, estimated to be the equivalent of 30 megatons of TNT, would have wiped out all of New York City and caused great destruction in neighboring Connecticut and New Jersey.

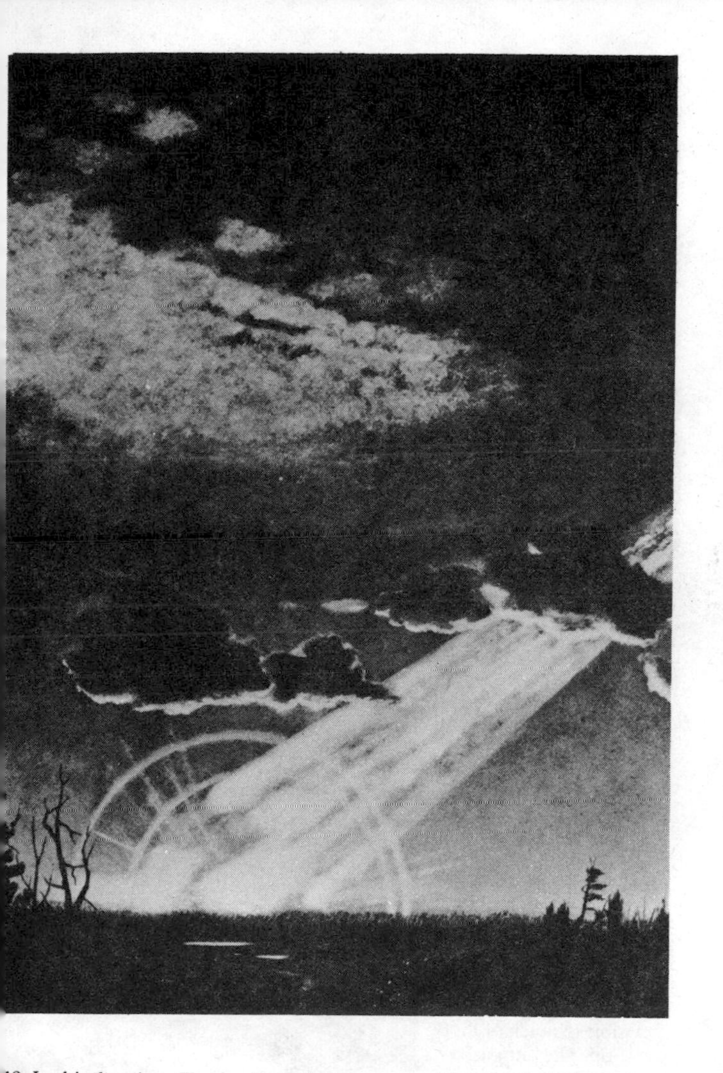

19. In this drawing, Chesley Bonestell shows the Siberian explosion as it might have appeared had it been caused by a small comet. The absence of any meteoritic craters or fragments led to the proposal of this idea as a possible solution to the mystery of the explosion, but the comet theory was eventually rejected by most experts because it did not match the physical evidence nor the testimony of eyewitnesses.

20. Halley's Comet, with its spectacular 90-million-mile tail, streaks across the night sky in 1910. No comet has ever collided with the earth, and no object remotely resembling such a "dirty snowball," as comets are sometimes called, was observed prior to the Siberian explosion. The cosmic body seen in 1908 over the Tunguska was described as a "bluish-white," cylindrically shaped object, streaming a fiery trail.

21. The Ikeya-Seki Comet of 1965 provided an example of the striking appearance of a comet in the sky against a background of stars. As it sped through our solar system, this sun-grazing comet was visible to much of the world. The Tunguska object, moving at a relatively slow velocity comparable to the speed of modern reconnaissance planes, was seen only during its passage over Siberia.

22. The flash of an atomic fireball as observed from the air only microseconds after the detonation of a test bomb over Bikini Lagoon on July 1, 1946. Eyewitnesses noted a similar blinding burst of light that "split the sky in two" over the Tunguska region on the morning of the 1908 explosion.

23. The fireball of an atomic blast detonated at an altitude of 750 feet at the Yucca Flat, Nevada, test site in 1957. The estimated height of the Tunguska explosion is two to three miles; its "pillar of fire" was visible for hundreds of miles.

24. A brilliant ionization glow surrounds the cooling fireball of the Diablo test at Yucca Flat, Nevada, in 1957. After the Tunguska explosion in 1908, ionization caused massive "silvery clouds" and lurid sunsets over a period of several days.

25. The mushroom cloud of the plutonium bomb, code-named "Fat Man," dropped on Nagasaki, Japan, on August 9, 1945, at the end of World War II. A similar mushroom-shaped cloud rose above the Tunguska region after the explosion of 1908.

26. Ground zero directly below the explosion of the atomic bomb over Hiroshima, Japan, August 6, 1945, at the end of World War II. The trees still standing around the Japanese Agricultural Exposition Hall bear a strong resemblance to the scorched but upright trees directly below the center of the Siberian explosion of 1908.

27. The destruction of Hiroshima was caused by a single uranium-235 bomb, code-named "Little Boy," dropped from the U.S. Superfortress *Enola Gay* on August 6, 1945. Soviet writer Aleksander Kazantsev was the first observer to comment on the extraordinary similarity between the patterns of destruction in the Japanese city and the charred, flattened Tunguska forest.

28. The radiant heat of the Hiroshima fireball burned the exposed paint, causing this permanent "shadow" of a hand-valve wheel which was more than a mile from ground zero. Of the Tunguska blast, one scientist noted that in a town 125 miles away "the flash of light from the explosion was so strong that it caused secondary shadows in rooms with northern exposures." The radiant heat caused intense pain to people 40 miles away in the Siberian trading station Vanavara.

29, 30. Hiroshima Castle, the headquarters of the 5th Division of the Japanese Army, before and after the atomic bomb exploded 1,800 feet directly overhead. The upright black and leafless trees around the castle were identical to those in the burned, defoliated zone in the heart of the Tunguska region.

31. Kulik and a sled husky stand before a section of "telegraph pole" forest resembling the charred, upright trees around Hiroshima Castle.

32. Later Soviet expeditions discovered in the Tunguska soil thousands of tiny magnetite globules formed by the heat of the explosion. These magnetized iron oxide spheres, perhaps of extraterrestrial origin, were believed by some analysts to be the only remaining fragments of the cosmic object that caused the blast.

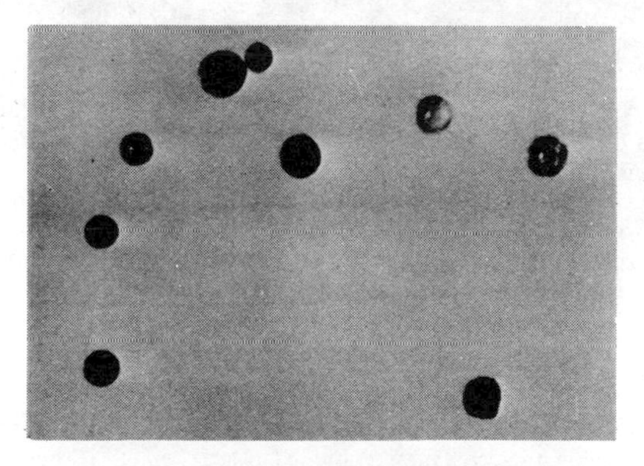

33. Also found in the Tunguska soil were numerous globules of silicate in small, glassy droplets or bulbs. In some instances, a particle of iron was found inside a silicate globule: the tremendous heat of the explosion had caused them to coalesce, like the congealed particles of trinitite found in 1945 at the atomic test site at Alamogordo, New Mexico. Many scientists were convinced that the particles imbedded in the Tunguska soil had not come from any known natural cosmic body.

34. This 1963 photograph shows a portion of the Tunguska region as it appeared fifty-five years after the blast, with new trees growing amid the dead, broken trunks and debris of the older forest. Later expeditions found, from examination of tree height and growth rings, that after 1908 the rate of growth in surviving trees as well as in trees which germinated after the explosion was markedly accelerated.

35. Aleksander Kazantsev, Soviet writer and space authority, first proposed in his story "A Guest from the Cosmos" that the Siberian explosion may have been caused by an extraterrestrial craft.

36. Felix Zigel, a professor at the Moscow Institute of Aviation is a leading figure in modern Soviet studies of the 1908 Tunguska blast. After re-examining the eyewitness and physical data, he and several other aerodynamics experts came up with estimates of the cosmic object's speed and flight path that differed radically from earlier estimates.

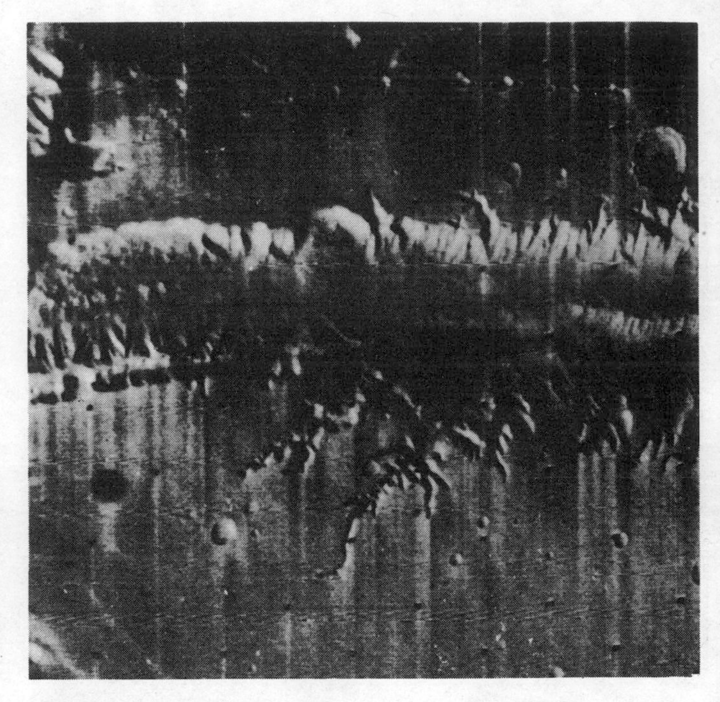

37. A photograph of the surface of Mars taken from an altitude of 1,225 miles above the planet on January 12, 1972, by NASA's Mariner 9, the first U.S. space probe to orbit another planet. Locked in an ice age, Mars has no water and is dotted with huge volcanoes and craters. Pictured here is an enormous valley, about 300 miles south of the Martian equator, with branching canyons extending into plateaus. Kazantsev conjectured that the cold, wind-swept Martian landscape was similar to the frozen, arctic tundra stretching just above the vast Siberian taiga. The Martians may have come to earth, according to Kazantsev's theory, to search for water for their dying planet.

38. Lake Baikal in southern Siberia, not far from the Stony Tunguska region. Over a mile deep in places and containing morc than a thousand species of plant and animal life found nowhere else on earth, Lake Baikal has an eerie reputation locally. Is it possible that an extraterrestrial craft was headed for the lake when it crashed?

39. An artist's conception of NASA's Pioneer 10 over Jupiter's "red spot." Launched on March 3, 1972, this unmanned space probe was the first to fly by Jupiter, a planet whose volume is one thousand times that of the earth and whose mass is more than twice that of all the other planets in our system combined. After passing Jupiter, Pioneer 10 became the first man-made object to leave our solar system for interstellar space. A plaque attached to the craft contained representations of human figures and a message designed for extraterrestrials.

Hersey describes the experience of one doctor who "was one step beyond an open window when the light of the bomb was reflected, like a giant photographic flash, in the corridor. He ducked down on one knee and said to himself . . . 'Sasaki, *gambare!* Be brave!' Just then (the building was 1,650 yards from the center), the blast ripped through the hospital. The glasses he was wearing flew off his face; the bottle of blood crashed against one wall; his Japanese slippers zipped out from under his feet—but otherwise, thanks to where he stood, he was untouched."

Into the vacuum created by the fireball was sucked all the debris, the smashed timber, the inflammable detritus of the shattered city. It fountained upwards in the mushroom cloud, 40,000 feet high and visible more than 400 miles out to sea. An official eyewitness report of the blast noted that the cloud "was observed from combat airplane 363 nautical miles away with the airplane at 25,000 feet. Observation was then limited by haze and not curvature of the earth." Of the blast it stated, "There were two distinct shocks felt in combat airplane similar in intensity to close flak bursts. Entire city except outermost ends of dock area was covered with a dark gray dust layer which joined the cloud column. It was extremely turbulent, with flashes of fire visible in the dust. Estimated diameter of this dust layer is at least 3 miles. One observer stated it looked as though whole town was being torn apart with columns of dust rising out of valleys approaching the town. Due to dust visual observation of structure damage could not be made." In a curiously appropriate note, the report concluded, ". . . other observers felt this strike was tremendous and awesome even in comparison with New Mexico test. Its effects may be attributed by the Japanese to a huge meteor."

To the Japanese, what had happened to Hiroshima was indeed as inexplicable as the impact of a meteor. Not even the massive chemical explosions of the heavy daylight raids that they had endured for weeks could compare either in form or intensity with this catastrophe. The sudden, chance phenomena characteristic of an atomic blast, with its multiple shock waves and thermal wave, astonished them. Men were knocked flat while those next to them remained upright. Others were stripped of their clothes but otherwise unharmed. Unlike an earthquake, the tremors resulting from the blast did not persist; yet the damage to the city was far greater than could have been caused by any natural disaster. Of the 75,000 houses in Hiroshima, 50,000 were totally and 18,000 partially destroyed. According to one report, "The destruction was total, regardless of the type of construction, within the half-mile radius. It was severe a little beyond the 1-mile radius. . . . From the 1 mile point out to 1 5/8 miles the damage was moderate. . . . From this point out up to 2 miles the damage was partial. Plaster was cracked and windows broken out to 8 miles from ground zero. The total damaged area in Hiroshima . . . was 18 square miles."

Later observers were impressed by the area's relative freedom from radiation; though in an air burst the gamma ray output is intense, the effect does not persist. The air burst therefore ensured that Hiroshima would suffer less from radioactive fallout. The experimental bomb at Alamogordo, New Mexico, which had been exploded on a hundred-foot steel tower, left the proving ground dangerously "hot." Dust and debris weighing thousands of tons, as well as the vaporized remains of the tower and a nearby farmhouse, were sucked up into the fireball, then sent billowing 40,000 feet into the air, where the mass flattened out in the soon-to-be-familiar mushroom-shaped cloud. This highly radioactive debris drifted across New Mexico, causing radiation blisters on cattle and forcing the Army to consider mass evacuations of some towns in the area. Another important factor discovered at Alamogordo was that the presence of dirt and other particles in the atomic fireball also impaired its efficiency. Consequently, the decision was made only days after the test blast that any bomb used against the Japanese would have to be air burst, thus guaranteeing the maximum possible damage from heat and concussion.

As terrifying as scientists found the New Mexico bomb's

explosive damage, they were more in awe of the terrible effects of radiation and heat. Heading towards the bomb site only a few hours after the blast in Sherman tanks sheathed with lead, the observers found an area for which they had no basis of comparison. Speaking of one observer, Lansing Lamont wrote in *Day of Trinity:*

> A half-mile away he saw what looked like a great jade blossom amid the coppery sands of the desert. Where the shot tower had once stood, a crater of green ceramic-like glass glistened in the sun. The fireball had sucked up the dirt, fused it virescent with its incredible heat, then dumped the congealing particles back on the explosion point. They lay there inside a 1200-foot-wide saucer some twenty-five feet deep at the center. The bomb, even from 100 feet up, had so pulverized the earth that the tower's concrete stumps, which once stood above the ground, had been crushed to a depth of seven feet beneath the sand. The tower itself had completely evaporated. Within a mile of the crater there was no sign of life or vegetation.

Scientists named the green glass "trinitite," after the Alamogordo code name "Trinity." Similar tiny globules of glass, many containing fragments of melted metal, were found in the soil at the site of the Tunguska blast. These globules appeared to have been formed by extraordinary heat.

Of the blast effect on flora and fauna in New Mexico, Lamont writes, "The stench of death clung to the desert in the vicinity of the detonation. No rattlesnake or lizard —nothing that could crawl or fly—was left. Here and there carbonized shadows of tiny animals had been etched in the hardpacked caliche, where the rampaging blaze had emulsified them. A herd of antelope that had been spotted the day before had vanished, bound, some said later, on a frightened dash that ended in Mexico. The yuccas and Joshua trees had disappeared in the heat storm; no solitary blade of grass was visible."

An "electromagnetic storm" which followed the blast

had deranged many of the delicate instruments; the earlier flash had penetrated shielded cameras, rendering them useless. The few samples of soil gathered by the shielded tank proved too "hot" to analyze, and radioactive fallout remained strong enough even a hundred miles away, where it settled in a white mist to blister the backs of some cattle and raise "gray spots" on the hides of others. The signs of the New Mexico test, because of the "dirty" fireball and resultant fallout, remained in evidence for some time.

Above the Alamogordo test site towered what was to become the most feared and potent symbol of the atomic age, a mushroom-shaped cloud. After the explosion, the still-expanding ball of lighter gases and dust, superheated beyond any fire that occurs naturally, erupted into the thin upper atmosphere about five miles above the earth; instantly the ball of gas was surrounded by a violet glow as heat and radiation ionized the air, creating a more vivid version of the aurora sometimes seen in the higher latitudes where changed particles from the earth's magnetic belts are drawn down to react with the atmosphere. Carried by the winds of the upper atmosphere, the glowing ionized air above the New Mexican desert traveled around the world creating, as did all later atomic explosions, atmospheric glows and odd, prolonged sunsets that faded within a few days as the air cooled and reverted to a more stable form.

Scientists in 1945 noted a substantial difference between the atomic cloud and the relatively familiar signs of Krakatoa in 1883. Funneled upward by the narrow venturi of the volcanic crater, such debris rises far higher than the relatively thin gaseous atomic cloud which is quickly dissipated by the first high-altitude winds. Dust from Krakatoa rose into the belt of fast jet streams that lie between 12 and 20 miles up—winds still known as "Krakatoa winds." Spread around the world by these high-speed air currents, volcanic dust forms a belt as much as three miles deep in the upper air. As the sun sets, light striking the underside of the cloud creates the crimson sunsets familiar from Krakatoa and such recent eruptions as that of Fuego volcano in Guatemala in 1975. Most brilliant about thirty minutes after the sun has sunk below the horizon, the volcanic sunset may persist for as long as a year, until the airborne dust gradually sinks to earth.

Part of the United States's justification for the attack on Hiroshima had been that the city held the headquarters of the 5th Division of the Japanese Army. Radio reports named Hiroshima as a "military center" on the basis of its presence there, and the castle in which the division was billeted became the official aiming point for the B-29 bombardier. Later, Air Force men remarked that the bomb had been dropped almost exactly on target. The whole unit was at calisthenics in the open air when the bomb fell; every member died instantly.

American physicist Philip Morrison, one of the observers sent into the city thirty-one days after the blast, was astonished at the relative lack of damage to the castle area, which must have been almost directly under the fireball. "Morrison passed the wreckage of the castle which had been the Fifth Division's headquarters," records Daniel Lang in *From Hiroshima to the Moon.* "The four thousand soldiers there had been killed. Morrison's guide lamented the destruction of the castle, part of which had been used as a military museum, where some souvenirs of an ancient victory had been preserved. The loss of these treasures distressed the guide. So did the fact that a tree planted by Hirohito's father had been burned black and leafless. Some water lilies in the moat of the castle had turned black, too, the guide added, but he was happy to say that they had begun to grow again. 'I wanted to make sure of that,' Morrison said, 'and I asked him to show me the lilies. They were growing all right.' "

The fact that the lilies were still growing proved to Morrison that, because the blast had been an air burst, the area had not been as completely saturated in gamma radiation as Alamogordo had been, where all life had been obliterated. The growth of the Hiroshima lilies was found to be characteristic of the accelerated plant growth which often followed atomic explosions. Not only did the black, leafless yet upright trees around the Hiroshima Castle at ground zero resemble the zone of burned, standing trees on the Tunguska site but a similar acceleration of growth was discovered in plants and trees closest to the center of the destroyed region in Siberia.

Just as the Japanese survivors first saw a "blinding flash" or "sheet of sun" in the sky, the Siberian witnesses recalled that the 1908 fire which suddenly appeared over the Tunguska was "brighter than the sun." The eyewitness descriptions of the Siberian explosion, in fact, coincide remarkably well with those of Hiroshima survivors. According to one account of the morning, "a huge flame shot up that cut the sky in two." Others spoke of an enormous "tongue of flame" or "pillar of fire" that flared over the taiga, followed by a tall, billowing column of "black smoke"—images that fit a nuclear fireball and mushroom cloud.

As at Hiroshima, the brilliant flash in the sky was accompanied by terrible heat. Sitting on his porch in Vanavara, 40 miles from the explosion, S. B. Semenov saw a "huge fireball" that gave off such a fierce heat that he could hardly remain seated. P. P. Kosolapov, a neighbor who was working at the side of the house, felt that his ears were scorched. The heat "seized" Semenov, nearly burning the shirt off his back. Then it grew dark, and he was thrown to the ground by the concussion. The shock wave, which immediately followed the flash and heat, "shook the whole house," damaging the ceiling and breaking windows. After the shock came the loud roar of the blast, like distant thunder.

In April of 1927 the Tungus Ilya Potapovich had informed Kulik that the "center of the firestorm" of the blast occurred at the pasture land of his relative Vasily Ilich, who had a herd of fifteen hundred reindeer. "In the same area he owned a number of storehouses where he kept clothes, household goods, reindeer harnesses," Kulik was told. "The fire came by and destroyed the forest, the reindeer, and the storehouses. Afterward, when the Tungus went in search of the herd, they found only charred reindeer carcasses. Nothing remained of the storehouses; everything had burned up and melted—clothes, utensils, reindeer harnesses, dishes, and samovars. Only a few pails remained."

No later investigator, not even those most completely committed to the meteorite theory, ever discredited these eyewitness reports. Krinov was initially perplexed by Semenov's description of the blaze of heat that seized him, since this was not consistent with a meteorite impact, but later he had to acknowledge that "Semenov's report about the sensation of 'heat' that occurred from the north side at the moment of explosion is of great interest. Apparently the 'heat' was a direct result of the radiant energy of the explosion. He also noticed that a 'hot wind' followed, which apparently was the explosive wave that had reached Vanavara."

Summing up the evidence of the flash and heat wave, Felix Zigel wrote in *Znaniye-Sila* in 1961 that in order for the explosion to have caused a sensation of burning 40 miles away in Vanavara, the energy must have been "not less than 0.6 calories per square centimeter." He pointed out that in the village of Keshma, 125 miles from the epicenter, "the flash of light from the explosion was so strong that it caused secondary shadows in rooms with northern exposures. . . . All these estimates are independent of one another and show one fact: the radiant energy of the Tunguska explosion comprised several tens of percent of the total energy. But this correlation between the parameters is characteristic only of nuclear explosions. A chemical explosion is excluded without a doubt—for a chemical explosion the ratio between these parameters is much less. Knowing the basic parameters of the Tunguska explosion, its temperature can be calculated. It turns out that the temperature was several tens of millions of degrees!"

Zigel evaluated the evidence of the burned trees, which were set on fire by the radiant heat within a radius of

8 to 10 miles. "Even L. A. Kulik noticed the unusual character of the burning of the trees in the region of the catastrophe," he stated. "According to the evidence of Kulik's closest colleague, E. L. Krinov, 'one can often see on a tree, particularly at its top, both thick and very slender branches side by side which have been broken off at the end and have bits of charcoal.' This is evidence that the burning was instantaneous, that is, as a result of explosion that followed and not from an ordinary forest fire."

Expeditions of the late 1950s and 1960s analyzed in detail the evidence of the burned forest. A. V. Zolotov, a prominent geophysicist who led 1959 and 1960 expeditions, remarked of the forest that "one observes an alternation of burned and unburned parts of the area and also an alternation of burned and unburned branches at the top of the same, completely burned tree." Zigel commented, "This means that the combustion of the trees was caused by light radiation from the explosion, and only in those places which were not in the shadow of leaves and branches, that is, it was a radiant burn."

Florensky, as violent an opponent of the atomic theory of the Tunguska explosion as Zolotov was its supporter, devoted much of his time on an expedition in 1961 trying to disprove Zolotov's theories, but the reports of the forestry experts he took with him agreed that "the 1908 fire flared up at several points" and that "the rounded shape of the fire site and the complete destruction by fire of the old forest over an extremely great area are outstanding features of the area; it differs in these respects from ordinary forest burn-out." Even Florensky admitted that "no one has dismissed the possibility of a flash fire," while hastily adding that "it has not been definitely established either."

Florensky's various expeditions, in 1958, 1961, and 1962, also completed the work of establishing the exact point at which the object had exploded. Had it indeed, as Kazantsev suggested, been an air burst? Both G. Plekhanov, of the Tomsk Medical Institute, and Zolotov, measuring the so-called "telegraph pole forest" and noting the way certain hillsides seemed to have been protected from the blast while others were entirely flattened, had concluded that only one explanation fitted these phenomena —an air burst about two miles above the taiga. In 1958, examining the barographic traces from 1908 still kept at the Meteorological Office in Potsdam, Germany, scientists decided that the object seemed to have blown up about two miles above the ground.

Other investigators set out to test the theory still further, sometimes with unconventional methods. In 1959 I. T. Zotkin and M. A. Tsikulin built a 1/10,000-scale model of the Tunguska plateau and studded a 7-by-10-foot construction with tiny pegs that would topple in the slightest breeze. Above the model they stretched a slanting wire with a tiny bomb attached. Meticulously setting it off at different altitudes, they found that when the charge went off at an angle between 27 and 30 degrees above the horizon, the pegs toppled in almost exactly the same pattern as the trees on the Tunguska. They differed with the barographic evidence only in the estimated height of the explosion; according to their calculations, five miles up was a closer estimate.

As a proponent of the comet collision theory, Florensky searched diligently for evidence to discredit this claim. But after detailed calculation he had to announce that "G. M. Zenkin places the emission center at 1.5 km to the southeast of the epicenter of destruction, at a height of about 5 km." The existence of a nuclear-type air burst over the taiga could no longer be denied.

What about the appearance and height of the other sign of an atomic blast—the mushroom-shaped cloud? I. S. Astapovich, correlating the results of eyewitness reports taken over a period in the area, had written in 1933 that "the explosion was observed from many points in the form of a vertical fountain," and he went on to list some of the extraordinarily consistent descriptions of what witnesses had seen. In Kirensk, it was "a fiery pillar . . . in the form of a spear," while to those at Nizhne-

Ilimsk it "changed into a fiery pillar and disappeared in a moment." All witnesses agreed on observing the rising "dense cloud" or "a huge cloud of black smoke," even though their homes were scattered over hundreds of miles. Astapovich calculated that at "a distance of 450 to 500 kilometres from the place of the explosion, the lowering of the horizon amounts to 16 or 20 kilometres. Hence it must be granted that both the pillar of the explosion and the 'black smoke' rose to a height considerably greater than 20 kilometres."

Twenty kilometers is about 66,000 feet. The Hiroshima cloud rose to between 40,000 and 50,000 feet. In a 1951 article containing further data on the Tunguska explosion, Astapovich concluded that "the cloud afterward was exactly like an atomic mushroom cloud."

Other eyewitnesses spoke of the "black rain" which fell following the Tunguska blast—a rain identical with that observed at Hiroshima. Astapovich also mentioned other phenomena associated with atomic explosions such as the atmospheric glow and strange sunsets seen for a few days after the 1908 explosion. These night displays persisted from June 30 to July 3 throughout Russia and Europe, and at the same time enormous "silvery clouds" were observed at great heights against the bright twilights.

In 1960 Plekhanov, a major figure in the Tunguska investigation, compared these midnight lights with those following the American nuclear explosion at Bikini Atoll in 1958. He found that almost identical, though smaller, atmospheric and magnetic effects followed Bikini as were observed in 1908. In a 1966 report, Soviet investigators V. K. Zhuravlev, D. V. Demin, and L. N. Demina determined that the magnetic and barographic effects, as well as the zone of destruction of the Siberian blast, corresponded with the 1958 high-altitude nuclear tests conducted by the United States. If the Tunguska event had taken place since 1958, they contended, it would have been immediately apparent to the new sophisticated instruments at the Irkutsk Observatory that a high-altitude atomic explosion had occurred.

The earlier meteorite and comet theorists found no explanation for the seismic shocks registered around the world in 1908—shocks never before noted in the case of a cosmic body striking the earth, but very similar to those recorded for atomic tests. Nor had meteors ever disturbed the earth's magnetic or gravitation field. Research in the mid-1960s by Sidney Chapman and Attia Ashour, published in the *Smithsonian Contributions to Astrophysics*, failed to find any but the tiniest electromagnetic disturbances caused by meteorites and none approaching the scale of those in 1908, which registered even on the primitive instruments of the time.

The most telling evidence for the atomic-blast theory is that of actual radiation damage to the area and to its flora and fauna. Much of the data collected by Tunguska observers and scientists on expeditions of the 1920s and 1930s, though little understood at the time, suggests that the Tunguska fall point was as much infected by gamma radiation as Hiroshima. In 1926, when the ethnographer I. M. Suslov questioned a gathering of 60 Tungus herdsmen at Strelka on their experience of the catastrophe, all agreed on the flare, the shock, and the resulting thunder—but also on an inexplicable aftereffect none had ever experienced before. The blast, Suslov said, "brought with it a disease for the reindeer, specifically scabs, that had never appeared before the fire came." These stories were "confirmed by everyone." Reports on the 1945 New Mexico tests record "gray patches" and blisters on the hides of cattle exposed to radiation.

After 1958, expeditions to the Tunguska region also noted genetic changes in the plants at the fall point since 1908. There was accelerated growth both in new trees and in those damaged by the explosion. In some cases, a fungus infection had spread over the dead wood, then been covered by the wild new growth. Examination of growth rings on living trees showed that around the period of the blast there was a noticeable increase in cell production; the rings were both wider and more pronounced. Those before 1908 varied from 0.4 to 2 millimeters in width, while after the blast they became as wide as 5 to 10 millimeters. Trees which germinated after the explosion would normally have grown to about 23 to 26 feet in height by 1958; instead, they towered between 55 and 72 feet high. Some of those which survived the blast were now almost four times their expected girth.

Zolotov's 1959 expedition drew attention to these anomalies, and Florensky's 1961 party made detailed examinations. The latter's report stated that "the features of accelerated tree growth established in 1958 have been confirmed by a great volume of data and are peculiar to the central region of the impact area. In view of the fact that the causes of the phenomena are not clear, work along this line should be continued." Later the report said, "Although literature sources indicate that the aftereffects of ordinary forest fires and forest uprooting with which we are familiar from European silviculture should not last longer than 15 to 20 years, they persist, occasionally without noticeable abatement, for a period of 40 to 50 years in the area of the meteorite fall. V. I. Nekrasov [participant in 1961 expedition] has expressed doubt as to the possibility of explaining this phenomenon in terms of purely ecological factors." Thinking that meteoric dust may have fertilized the area so as to encourage growth, the 1961 party planted test areas of grain to see if it grew better than in traditional soil. It did not, proving that whatever affected the trees at the fall point did so only at the instant of the blast and for a short time thereafter. The similarity to Hiroshima's water lilies is undeniable.

Finally, and most convincingly, a group from Tomsk led by Plekhanov in 1959 spent six weeks accumulating elaborate soil and plant samples. Three hundred samples of soil from miles around the blast area, eighty samples of ash, and one hundred plants were collected, each plant having as many as fifteen different specimens taken from its outer bark, its leaves, and its inner rings. A particularly sensitive spectrometer was used to examine the radiation level of all the samples. Even though fallout from Soviet H-bomb tests above the Arctic Circle in Siberia may have masked some radiation, the scientists found that "in the center of the catastrophe, radioactivity is one and a half to two times higher than along the 30 or 40 kilometers away from the center." As for the plants, it was true that the outer rings did show signs of radiation from later fallout, but within the inner rings, particularly in those surrounding the area of 1908, radioactive cesium 137 was found in quantities far above normal.

As the Tungus had reported, "the fire came by" in 1908. There could now be no doubt that this fire was atomic.

How big was this fire? Though the destructive power of the Hiroshima blast was awesome, the force of the Siberian explosion was far greater. In the Japanese city, wood was ignited by the blast at a distance of one mile, on the Tunguska plateau at a distance of 8 to 10 miles from the fall point. The destroyed area in Hiroshima totaled 18 square miles, that on the Tunguska 1,200 square miles. Japanese naval students felt a hot breeze from the Hiroshima blast 60 miles away, while 375 miles from the Tunguska blast, in Kansk—six times as far—a hurricanelike gust of air was experienced. At Hiroshima, severe heat was felt between 3 and 4 miles away; on the Tunguska about 40 miles away.

The atomic blast on the Tunguska was, at the very least, ten times more powerful than that at Hiroshima. But when one takes into account the greater altitude of the Siberian blast and the much larger total of forest destruction, the possibility exists of an explosion a hundred times more massive—in the megaton range. Soviet authorities and American experts such as the Nobel prize-winning chemist Willard F. Libby have estimated its energy yield as high as 30 megatons—fifteen hundred times as great as that at Hiroshima.

Nine

ANTIMATTER OR BLACK HOLE?

The overwhelming evidence of an atomic explosion in Siberia in 1908 prompted some scientists to look to rarefied areas of theoretical physics and astronomy for a natural explanation. If such an explosion could not be accounted for by any ordinary cosmic object, they asked, then could a previously unknown, extremely dangerous form of extraterrestrial matter have penetrated our atmosphere, producing a detonation equal to 10^{23} ergs of nuclear energy and yet leaving no physical trace of itself?

"It is clear that the Tungus cosmic body . . . could not have been a comet," wrote the geophysicist A. V. Zolotov, speaking for many of his fellow Soviet scientists. "Neither could it have been a normal ice, stone, or iron meteorite. The Tungus body obviously represents a new yet unknown, much more complicated phenomenon of nature than has been encountered up to this time."

Space expert Felix Zigel commented, "In spite of the great advances in our knowledge of the structure of matter, we are far from knowing all about the internal, 'deep' properties of matter, about the conditions under which nuclear energy can be released. We do not know but what on June 30, 1908, the earth collided with some very extraordinarily, still unfamiliar but natural heavenly body."

Could the great blast have been caused, for instance, by a small body of antimatter? Or perhaps a so-called "black hole" was responsible for the massive destruction of the taiga? Modern theoreticians examined both these newly conjectured cosmic phenomena as possible solutions to the Siberian riddle.

Even before the first experiments partially confirming the existence of antimatter had been conducted with the powerful atom-smashing accelerators at Berkeley and other scientific institutes, its presence in space had been surmised in the 1930s by Nobel prize-winning physicist P. A. M. Dirac and later imagined by science fiction writers like Jack Williamson who, in his *Seetee Shock*, utilized it as an element in a profound cosmological drama. The premise of antimatter is simple and logical. Why, physicists ask, should there not be floating in free space atoms in which the electrical charges of the particles are reversed—atoms in which positively charged particles (positrons) revolve around a negatively charged nucleus, rather than negative particles revolving around a positive nucleus, as in normal terrene atoms? If a fragment of antimatter came into contact with a terrene object, both would be instantly and totally annihilated.

In the view of some physicists and astronomers, antimatter might offer a natural explanation for scores of inexplicable phenomena.

In deserts around the world, for example, large pieces of fused greenish-yellow glass have been discovered that are almost identical with those found at meteor fall points. Samples have been collected in Libya, Australia, and across central and southern Africa. But no trace has been found of the meteorites that might have created such deposits.

Along the coastal plain of South Carolina, and elsewhere on the eastern seaboard, are found thousands of shallow, egg-shaped depressions known as "Carolina bays." From the air these bays are often quite apparent. "Geomorphologists," the Encyclopaedia Britannica notes, "have discovered no satisfactory explanation for these curious natural features."

To the north in Virginia, at the heart of the Great Dismal Swamp, Lake Drummond lies in a hollow, shallow and egg-shaped, burned out of the peat which forms the swamp. Indians claim a "firebird" created the depression, and scientists have suggested a meteorite as the cause of the burned area, which extends down through the peat to the sandy floor of the lake, but no meteoritic object had been found there.

On September 15, 1940, the New York *Times* reported, "As the 22-foot cutter-type sailboat *Rockit II* was crossing Long Island Sound near Bridgeport, Connecticut, yesterday morning with four peaceful persons aboard, a shell screeched across her bow and exploded in the water a hundred yards away." The passengers recalled, "The screech came first—an unholy noise. Then, a split second later, the explosion blew up a great tower of water, twenty or thirty feet in the air. It was the strangest thing in the middle of that peaceful Sound. Why, there wasn't even a boat in sight. And not an airplane overhead!" Later investigation also showed that no artillery shell could have exploded near the boat.

All these incidents were recorded in technical journals of meteor science as possible results of minute scraps of antimatter striking the earth. A small contraterrene meteor would create an explosion out of all proportion to its size, then vaporize, leaving no trace except possibly the enigmatic craters called Carolina bays, deposits of fused sand in the desert, or such mysterious happenings as the Long Island Sound phenomenon. It might also, some scientists pointed out, provide a plausible explanation for the events on the Stony Tunguska.

Initially put forward in a paper published in the February 1941 edition of *Contributions of the Society for*

Research on Meteorites, one of the most respected international journals of meteor science, the antimatter theory drew wide attention and some disagreement. Lincoln La Paz, author of the paper, then in the Department of Mathematics of Ohio State University and later a leading meteorite expert at various American universities as well as cotranslator into English of many papers by Leonid Kulik on the Tunguska phenomena, had anticipated the most common objections to the theory—that antimatter would explode on its first contact with the earth's atmosphere—by quoting physicist V. Rojansky's calculations that "an approximately cylindrical, contraterrene iron meteorite, falling with its axis vertical, will survive transit through the atmosphere. . . . If a contraterrene iron meteorite of a size comparable to those of the largest irons conjectured to have fallen should strike the earth, an extremely powerful explosion would result, since, in addition to the large store of heat energy resulting from the transformation of the kinetic energy of motion of the meteoric mass, a vast amount of energy would be liberated by its annihilation."

Willard Libby, the American chemist who had developed the carbon-14 dating technique, published an essay in 1965 with Clyde Cowan and C. R. Atluri arguing in favor of the "antirock" theory. "In searching for other natural means by which a large nuclear energy yield might be obtained," the article stated, "we are unable to find one other than the annihilation of charge-conjugate ('anti-') matter with the gases of the atmosphere." Speculating about the flight of an antimatter bolide through the air, they contended that "only a small fraction of the bolide could annihilate in flight" and that it might remain "essentially solid" until it came in contact with the heavier lower atmosphere where "continued annihilation might heat it to the gaseous stage and dissemble it explosively."

One result of this explosion would be the increase of radioactive carbon in the atmosphere. Estimating that the carbon-14 yield produced by the annihilation of even a small antirock might be comparable to the amount released in the atmosphere by later nuclear bomb tests, such as those at the U.S.S.R. test site at Novaya Zemlya, Libby and the other scientists measured deposits in American tree rings and found that the amount of radiocarbon increased after 1908, though they admitted that there were "uncertainties" in such proof.

The science-fiction writer Kazantsev also adopted the contraterrene theory as a possibility, since there was no reason why his Martian visitors should not have used antimatter as a component of their ship or its engine. But most Soviet scientists rejected the antimatter concept, arguing that contraterrene meteorites, even if they existed, could not explain the actual physical effects of the Tunguska blast.

The 1969 U. S.-sponsored committee report on unidentified flying objects, prepared under the direction of the physicist Edward U. Condon, examined the theory and pointed out that an antimatter explosion "has measurable consequences. When matter and antimatter come into contact, they annihilate each other, and produce gamma rays, kaons, and pions. If an antimatter meteoroid were to collide with the atmosphere, negative pions would be produced. The nuclei of the surrounding air atoms would absorb the negative pions and release neutrons. Nitrogen nuclei would capture the neutrons and be turned into radioactive carbon 14. As carbon dioxide, the radiocarbon would be dispersed throughout the atmosphere and be absorbed by living organisms."

The same report continued, "The energy of the Tunguska bolide was estimated from a study of the destruction that occurred. The initial quantity of antimatter and the amount of radioactive carbon dioxide produced was then estimated. Sections of trees which grew in 1908 were analyzed for radiocarbon. The conclusion of several scientists is that the Tunguska meteor was probably not composed of antimatter."

But to other scientists in the West, largely unfamiliar with the vast amount of data accumulated by later Soviet researchers, antimatter provided one of the few adequate explanations of the Tunguska explosions. This theory, however, was soon rivaled by another even more bizarre suggestion—that the explosion was the result of a collision between the earth and a "black hole."

As early as 1939 J. Robert Oppenheimer, a leading figure in the Manhattan Project that constructed the first atomic bomb, had speculated about other states of matter created by the pressures and temperatures of a collapsing star.

Most average stars, including those the size of the sun, eventually fade, gutter, and die like a bonfire, the outer layers collapsing on the dying core until it becomes a dense, spinning ball of neutrons. Was the same true of enormous stars, many times the size of our sun, scattered through the universe? Oppenheimer believed that these larger stars would collapse in an entirely different and terrifying way. As the outer layers fell inwards, the whole sun would become so dense as to form a new kind of matter, popularly known as a "black hole." Even a speck of this matter might weigh millions of tons.

British physicist John G. Taylor labeled a black hole the "supreme bizarre object" and compared it to a "cannibal, swallowing up everything that gets in its way. Once engorged by it, there is no hope of escape." A report in the London *Sunday Times* by journalist Tony Osman suggested that a black hole may be visualized as a "cosmic vacuum cleaner, sucking in stars, light, and anything else that comes within range. But it is more than that. A black hole is so dense that none of the laws of physics as we know them can apply and nobody can yet begin thinking of what laws do apply. Inside a black hole you have the origin of the universe run backwards."

Such objects would distort the fabric of space, sucking in light rays so greedily that they themselves and even the space around them became invisible; in the presence of another star they would draw in gas and emit floods of hard radiation. One estimate in the 1970s placed the number of black holes in the universe as high as a billion, ranging from some larger than our sun to others, perhaps formed at the beginning of the universe, with diameters no larger than that of a dust speck.

Recently, by means of a man-made satellite orbiting the earth, astronomers have detected X-rays emitted from the constellation Cygnus; these emissions indicate that in one of its binary stars, Cyg X-1, a blue supergiant appears to be circling an invisible star many times bigger than our sun. This invisible neighbor, feeding on the energy of the giant, may be a black hole.

As New York *Times* science editor Walter Sullivan has noted, "Nothing in the art of the medieval alchemist or the contemporary science-fiction writer is more bizarre than the concept of the black hole." In a recent article he examined the notion that "a tiny 'black hole' hit Siberia, passing through the entire earth and emerging in the North Atlantic." This idea was proposed in 1973 by A. A. Jackson and Michael P. Ryan, scientists at the University of Texas, to explain the blast in the Tunguska region. If such compressed "mini" black holes exist, one might have struck the earth, creating an affect akin to a nuclear explosion, then passing through the planet like a bullet until it exited on the other side and continued its rampage through the universe.

Soviet experts on the Tunguska event examined these new theories, but ultimately rejected them on the basis that they did not match the actual evidence. The enormous body of eyewitness testimony, supplemented by the findings of various expeditions, eliminate the possibility either of a black hole or a contraterrene particle.

Unaffected by the friction of the earth's atmosphere, a black hole probably would have struck without warning and left a deep crater heavily impregnated with hard radiation. The absence of a crater of any kind or severe hard radiation argues against this theory, as does the shape and entry speed of the 1908 object.

From the numerous witnesses who actually saw the object before it exploded in the sky, we know that it was of considerable size and had a markedly "cylindrical" shape, like a pipe or a tube. Some described it as resembling a "chimney." This "elongated flaming object" glowed with a "bluish-white" radiance brighter than the sun and left a broad trail of multicolored smoke in the atmosphere. In its descent over the Tunguska region, the object created a huge ballistic wave that was, according to experts, exactly the same as the air wave of a missile.

The velocity of this cylindrical, missilelike object was at first assumed to be as much as 30 or 40 miles per second, chiefly in order to account for the higher kinetic energy of the blast; but the Soviet geophysicist Zolotov later made a more accurate determination of the object's speed. From comparisons of the effect of the ballistic wave and blast force on trees in the region, he calculated that shortly before the explosion the velocity was probably not more than 1½ to 2 miles per second—about 7,000 miles per hour.

Professor Zigel points out that eyewitnesses saw the object overhead and heard its deafening roar simultaneously, which would have been possible only if the Tunguska object's velocity was just slightly greater than the speed of sound, or about .18 miles per second. If the object had passed through the atmosphere at a speed as fast as 30 to 40 miles per second, Zigel notes that "the eyewitnesses would have seen it first and only later would have heard the noise, somewhat as thunder is heard after the lightning has flashed." The final speed could not have exceeded a few miles per second, he concludes, or it would have been virtually impossible for witnesses to gain concrete visual impressions of the object's shape.

Neither the antimatter nor the black hole thesis could account for the slowly descending tube-shaped object streaming a fiery trail, the barrage of widely spaced explosions, the oddly shaped pattern of leveled trees, the sudden growth of vegetation after the blast, the clear evidence of an explosion above the taiga rather than on its surface, or the scores of other inconsistencies which had earlier discredited the meteorite and comet theories. The known facts seemed to thwart all attempts by scientists to explain the Tunguska occurrence by means of any natural phenomena.

Ten

THE CYLINDRICAL OBJECT

Is it possible that the flaming "cylindrical" object that exploded in mid-air over central Siberia in 1908 was, in fact, a spacecraft?

As the debate over the multimegaton blast continued through the sixties and into the seventies, an increasing number of authorities came to the opinion that the answer inevitably must be "yes"—partly because all natural explanations of the "cosmic body" seemed to wither under rigorous analysis but also because, as mankind entered the age of space technology, the idea began to appear more feasible. The belief that the object was artificial, moreover, gained new credence from findings based on the peculiar shape of the explosion and a study of particles found at the blast site, as well as some surprising recent calculations by aerodynamics experts about the flight path through the atmosphere.

The strangely irregular boundaries of the blast region had already been noted by E. L. Krinov during the third

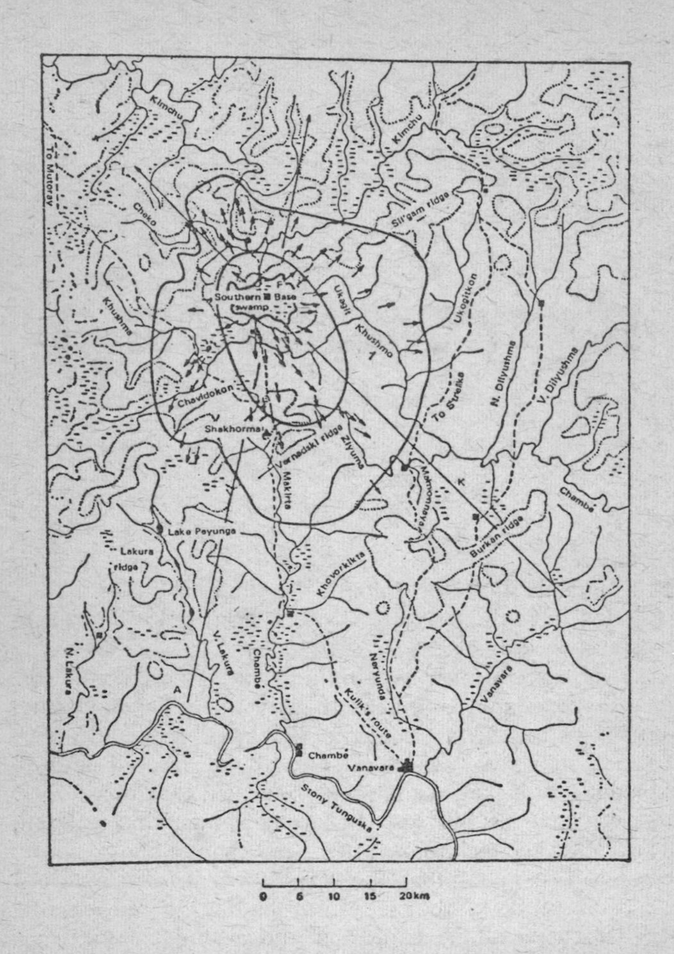

A 1958 map of the Tunguska region summarizing most of the data available then about the blast. The shape of the explosion wave and two different trajectories are shown, according to (A) I. S. Astapovich and (K) E. L. Krinov. (Courtesy of Pergamon Press)

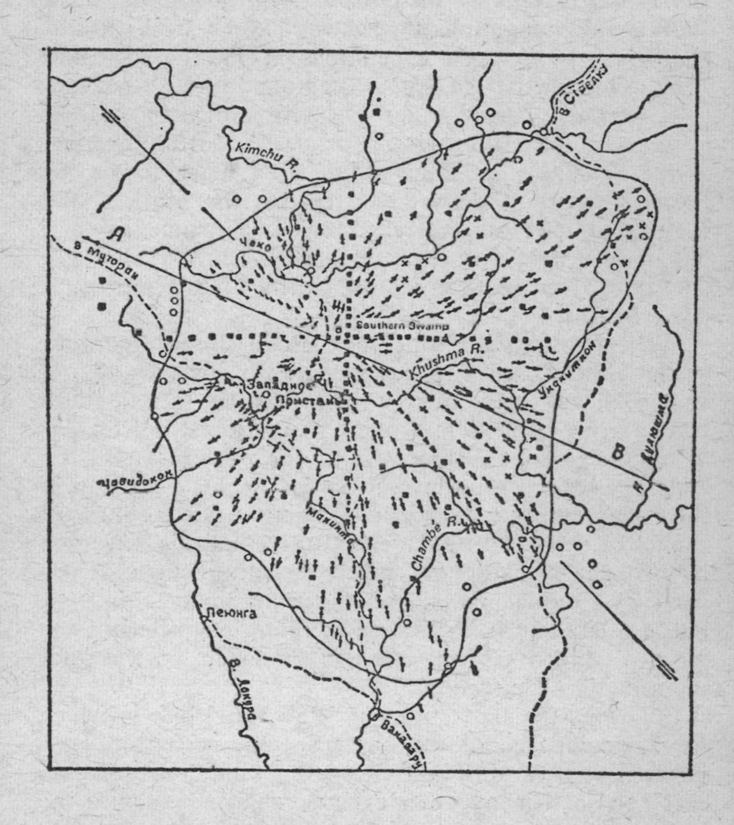

This 1961 chart, with arrows indicating the direction of the fallen trees, shows the peculiar oval or elliptical shape of the Tunguska blast pattern which puzzled early investigators. The map plots a trajectory approaching from east-southeast (line AB), with the heart of the destruction located in a slightly off-center position in the massive explosion wave.

Tunguska expedition in 1929. "The area of uprooted forest," he observed, "has an oval form with the major axis situated in a direction from southeast toward northwest." The oval shape surprised early researchers, who had expected the area of destruction to be circular. The 1938 aerial photographs further verified this oval pattern; and K. P. Florensky's expeditions in 1958, 1961, and 1962 determined from extensive ground and air surveys that the 1,200 square miles of forest leveled by the explosion and the scattering of "cosmic dust" from the blast had a definite elliptical contour. New maps drawn by Florensky's group showed clearly that the heart of the destruction, the totally scorched region containing the dead but upright trees, lay in a slightly off-center position in an explosive wave that fanned chiefly toward the south and northeast.

What was the significance of the odd elliptical contour of the blast? Discussing this issue, Felix Zigel wrote, "It is very evident on the map of the region that the boundary of the area of complete leveling of the forest is irregular in outline. Also, the epicenter of the explosion and the zone of trees left standing occupy an eccentric position in the region of the catastrophe. Obviously, this asymmetry cannot be explained by the effect of the ballistic wave due to the flight of the body—the zone of destruction is elongated in a direction that is not parallel to the trajectory but at a large angle to it." He therefore characterizes the blast as "directive"—the effect of the explosion was "not the same in all directions."

Following his expeditions in 1959 and 1960, in which he reexamined all of the physical evidence in the Tunguska region, A. V. Zolotov came up with an explanation that Zigel and many other experts found acceptable: the blast had an unusual oval shape because the explosive material was encased in some type of "container." The structure of this container, like the thick paper cylinder of a large firecracker, caused the explosive charge to fan out elliptically as it burst. "The directivity of the explosion," Zigel commented, "was due to the inhomogeneity of the container." The Tunguska object "consisted of at least two parts: a substance capable of nuclear explosion and a nonexplosive shell."

But is there any concrete proof of such a nonexplosive container? Some analysts believed that at least partial proof had already been acquired by Kulik—in the soil samples brought back by his expeditions. In the late fifties, when these specimens were subjected to extreme magnification and careful laboratory testing, small particles of extraterrestrial matter were discovered. The Tunguska soil contained concentrations of spherical globules, a few millimeters or less in size, composed primarily of silicate and magnetite, a magnetized iron oxide. The magnetic globules resembled little droplets or bulbs and were sometimes linked together in clusters. Krinov, who took part in the soil study (which had been conducted by A. A. Yavnel), notes that "there are even instances where a magnetite globule . . . was discovered in a completely transparent silicate globule." These separate particles appeared to have "coalesced" under tremendous heat, like the trinitite found at the Alamogordo atomic test site.

If the spheres were not of terrestrial origin, were they simply the usual micrometeorite dust that falls daily across the entire surface of the planet or could they be fragments of the Tunguska object? In 1962 Florensky attempted to resolve these questions. Using a helicopter, his team charted the pattern of the explosion's scattering ellipse over a large area and then collected a wide range of samples for chemical analysis. "At the expedition's base camp in the taiga," he wrote, "we set up a concentration mill, to separate an infinitesimal admixture of extraterrestrial matter from numerous bags of soil samples. Gradually the picture began to take shape, as the outlines of the dispersal pattern became more and more distinct." As they had expected, thousands of "tiny brilliant spheres," many fused together, were found imbedded like pellets in the earth and trees. The pattern of their distribution, moreover, conformed to the elliptical blast wave; the particles appeared to have been dispersed over the Tunguska territory by the "updraft of heated air" from the explosion. Florensky, still defending the obsolete comet theory, maintained that they were molten cometary debris; but Zolotov and many others were certain that the fused particles could not possibly have come from any known comet or meteoritic body.

This notion was strengthened when more detailed analysis of the spheres revealed small amounts of cobalt and nickel, as well as traces of copper and germanium. Kazantsev and those supporting his views argued that the discovery of these metallic elements offered further proof of an artificial craft. "Remember that the ship must have had electrical and technical instruments," Kazantsev elaborated, "also copper wires, and surely means of communication—semiconductors containing germanium." The exact source of the strange globules, however, has not been fully determined. It is apparent that the congealed particles resulted from the enormous heat of the explosion; and it is possible to surmise, with a strong degree of probability, that they may indeed be the only existing remnants of the cylindrical object—the "nonexplosive shell" that housed the atomic fire.

Soon Zigel and several other aerodynamics authorities, experienced in modern rocket technology and upper atmospheric trajectories, came forward with some astonishing assertions that ultimately tipped the scales in favor of a spacecraft. When A. Y. Manotskov, an airplane designer, mapped the passage of the Tunguska object through the air, his calculations agreed with Zolotov's findings that the object must have arrived at a velocity much slower than that of a natural cosmic body. Meteorites usually rush into our atmosphere at speeds of between 9 and 13 miles per second, and sometimes as fast as 25 miles per second. Manotskov decided that the 1908 object, on the other hand, had a far slower entry speed and that, nearing the earth, it reduced its speed to "0.7 kilometers per second, or 2,400 kilometers per hour"—less than half a mile per second, which is comparable to the velocity of a high-altitude reconnaissance plane. Boris Liapunov, a Soviet rocket specialist, examined these estimates and concurred with Manotskov that the object had behaved in its entry and velocity like a supersonic craft.

What flight path did this craft follow through the earth's atmosphere? This has always been one of the most debated aspects of the Tunguska object. The principal clues to its trajectory are the observations of eyewitnesses and the ballistic shock damage caused by the rapid compression of air ahead of the moving object. A high-speed craft, such as a guided missile or supersonic spy plane like the U. S. Air Force's SR-71 which can travel faster than 2,000 miles per hour, gathers before its nose a conical wave of compressed air molecules that causes loud sonic booms when the waves reach the ground and, even at altitudes of several miles, can inflict damage at ground level. The deafening roar heard in June 1908 by witnesses throughout central Siberia during the flight of the Tunguska object was probably caused by its powerful ballistic wave; the series of thunderclaps heard afterwards resulted from the massive blast waves.

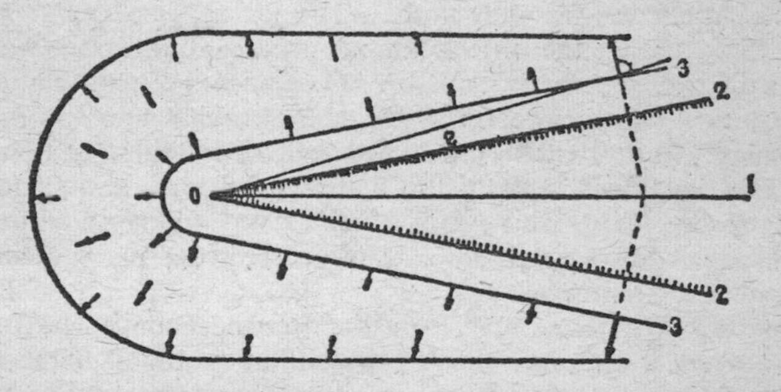

According to experts, the Tunguska object created a strong ballistic wave during its flight that was exactly like the shock wave of a missile. This Soviet chart depicts (1) the trajectory of the object (seen from overhead); (2) the ballistic wave at the instant of the explosion; (3) the ballistic wave after the explosion; and the angle of the tree fall from the blast's epicenter.

Most investigators have agreed that determining the flight trajectory was essential to uncovering the object's true identity. After studying the damage of the ballistic shock and the explosion itself, various investigating teams from the later 1920s to the present have arrived at very different conclusions about the flight path.

Three of the first interpreters of the Tunguska event, Voznesensky, Suslov, and Astapovich, basing their opinions primarily on eyewitness reports and seismic data, decided that the object moved from south-southwest to north-northeast. Kulik, after examining the blast effect around the Southern Swamp, believed that "the meteorite flew in the general direction from south to north," while Krinov proposed that "the trajectory ran from southeast to northwest" and placed the "projected position on to the earth's surface of the beginning of the trajectory as corresponding approximately to the northern shore of Lake Baikal."

Later, Florensky's findings convinced him that "both the general pattern of the toppled trees and the relationship between the centers of fallen deadwood and the searing effects—as well as the distribution of cosmic dust —indicate that the object came from east-southeast." According to another scientist, V. G. Konenkin, east-southeast to west-northwest was the most probable trajectory. In his field investigations, Zolotov examined standing trees that bore traces both of the ballistic and blast shock; he concluded that the air wave, which caused relatively minor damage compared to the explosion, had definitely come from the southwest.

Did the object arrive from the southeast or the southwest? At first the problem appeared impossible to resolve, for eyewitness testimony and forest damage could be produced to support either direction. The object had been visible overhead as a "fiery body" to villages near Kansk, southwest of the blast, but it had also been seen in Kirensk and other towns lying to the southeast. Scores of separate, reliable observations made both flight paths seem equally feasible, though obviously the same object could not have

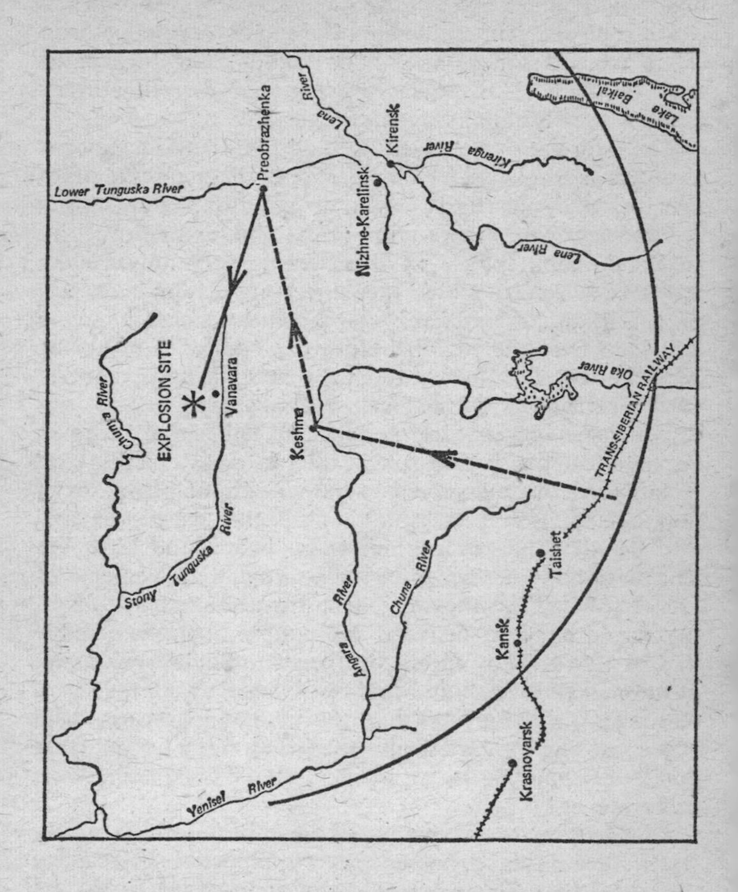

The flight path of the cosmic object, as reconstructed from eyewitness testimony and ballistic wave evidence. Felix Zigel and other space experts agree that, prior to exploding, the object changed from an eastward to a westward direction over the Stony Tunguska region. The arc at the bottom of the map indicates the scope of the area where witnesses either saw the fiery object or heard the blast.

appeared almost simultaneously in two different locations hundreds of miles apart.

Or could it? The problem of the different trajectories was eventually solved with a startling answer: *both* paths were accurate. The object had switched direction in its journey over Siberia!

The information acquired by the Florensky and Zolotov expeditions about the ballistic shock effect on the trees provides a strong basis, in some scientists' view, for a reconstruction of an alteration in the object's line of flight. In the terminal phase of its descent, according to the most recent speculations, the object appears to have approached on an eastward course, then changed course westward over the region before exploding. The ballistic wave evidence, in fact, indicates that some type of flight correction was performed in the atmosphere.

The same opinion was reached by Felix Zigel, who as an aerodynamics professor at the Moscow Institute of Aviation has been involved in the training of many Soviet cosmonauts. His latest study of all the eyewitness and physical data convinced him that "before the blast the Tunguska body described in the atmosphere a tremendous arc of about 375 miles in extent (in azimuth)"—that is, it "carried out a maneuver." No natural object is capable of such a feat. Thus Zigel, together with Soviet rocket and aviation experts such as Manotskov and Liapunov, joined Kazantsev in believing that the remarkable cylindrical object causing an ellipitically-shaped atomic blast in 1908 could only have been "an artificial flying craft from some other planet."

In addition to its maneuvers near the earth's surface, the craft must have steered, as it approached from outer space, into a trajectory angle almost identical to the reentry path used by modern space vehicles. To make a successful entry through the dense air blanket surrounding our planet, a spacecraft must maintain a precise flight angle of minus 6.2 degrees to the horizon. American astronaut Michael Collins has described the difficulty of navigating back into the earth's atmosphere: "The allowable limits were spectacularly small. On the return trip, the atmospheric 're-entry corridor,' or zone of survivability . . . was only forty miles thick, and hitting a forty-mile target from 230,000 miles is like trying to split a human hair with a razor blade thrown from a distance of twenty feet." Missing this corridor can be fatal, for if the angle is too steep, the craft will burn up in the severe frictional heat; if too shallow, it will rebound back into space. In order to penetrate to the lower layers of the earth's atmosphere, the craft of 1908 must have executed the hazardous trip through the entry corridor with near perfection.

According to the Soviet astronomer B. Y. Levin, who made a careful study of the Tunguska object's movement through the atmosphere, the estimated height at which it was first sighted was approximately 80 miles. Witnesses saw a fiery tube "shining very brightly with a bluish-white light." The entry zone at the edge of the atmosphere is encountered at this same altitude; passing through this upper air layer, any sizable object, such as a spacecraft, acquires a luminous plasma coating that causes it to glow like a meteor with a brilliant white-hot radiance.

There is strong evidence, therefore, to indicate that an extraterrestrial vehicle may explain the Siberian explosion. Though the proof is not complete and cannot be viewed as absolutely conclusive, the discoveries of modern astrophysics and astronomy, particularly recent insights regarding the possibility of life existing elsewhere in the universe, are providing a new perspective in which this theory appears increasingly acceptable and logical.

Eleven

A COSMIC VISITOR

The Siberian superblast is approaching its seventieth anniversary. The scientist who first explored the immense destruction site but could not elucidate it, Leonid Kulik, has been dead for more than three decades; and most of the next generation of investigators who tried to make sense of the cataclysmic event are well into middle age or beyond. Aleksander Kazantsev, first to suggest that the Tunguska object was an interplanetary vehicle, lives in an apartment in Russia's capital. Not far from Kazantsev's home, a new generation of ambitious young researchers at Moscow University and the Academy of Sciences study the different theories about the event and sift through the factual evidence, hunting for the sure proof, the ultimate clue to the riddle that confounded their predecessors.

The amount of material accumulated on the subject in the Soviet Union is staggering; few events in Russian history have aroused as much sustained interest and controversy or have so irresistibly fascinated both the scientific community and the general public as the Tunguska "marvel." A bibliography printed in 1969 in the journal *Priroda (Nature)* listed 180 scientific papers, 940 articles, and 60 novels, as well as scores of stories, poems, films, and TV programs about the great Siberian explosion.

Nor have Western writers on technology and astronomy escaped the fascination of the day the fire came by in Siberia. Both the American Isaac Asimov and the Englishman Arthur C. Clarke, for instance, have pondered the drastic effects that would have occurred if the Tunguska object had exploded over a heavily populated city.

"A fall like that in the middle of Manhattan," Asimov wrote, "would probably knock down every building on the island and large numbers across the rivers on either side, killing several million people within minutes of impact."

Clarke, among the proponents of the meteorite theory, began his novel *Rendezvous with Rama* with a dramatic warning:

> Sooner or later, it was bound to happen. On June 30, 1908, Moscow escaped destruction by three hours and four thousand kilometers—a margin invisibly small by the standards of the universe. On February 12, 1947, another Russian city had a still narrower escape, when the second great meteorite of the twentieth century detonated less than four hundred kilometers from Vladivostok, with an explosion rivaling that of the newly invented uranium bomb.

Set in the twenty-first century, Clarke's novel focuses on a cosmic intruder entering our solar system, initially mistaken for an asteroid but eventually discovered to be an interstellar craft resembling a gigantic "boiler" more than thirty miles in length.

Despite all attempts to label the 1908 intruder from the cosmos with an indisputable identity, it remains after nearly seventy years one of the most intriguing mysteries of modern science. In 1960 Albert Parry, an expert on

Russian technology, commented: "Looking for incontrovertible proof that the messenger from the sky of 1908 was a spaceship and not a mere chunk of stone and metal is fast becoming a favorite sport for many adventurous souls in Soviet Russia. Newer and newer expeditions are going into the Tungus wilds to try for replies to baffling points now raised not only by Kazantsev but also by Zigel and his other supporters rallying to his romantic cause."

But now, in the second decade of the space age, it is no longer regarded as merely "romantic" to seek proof of the explosion of an extraterrestrial vehicle over Siberia. Scientific attitudes toward the possibility of intelligent life existing on other planets have radically changed in the past ten years. "Should this [the spacecraft theory] be finally confirmed by investigations now in progress," Zigel remarked, "the significance of the Tunguska disaster would be inestimable." Many experts today concur with this opinion and feel that it is not based on a whimsical hope.

Yet it is not likely that positive confirmation will result from new expeditions to Siberia. Although specialists from the Soviet Union and many other nations undoubtedly will continue to travel to the blast region to examine the scorched and broken trees—despite the passage of years, the original physical appearance of the shattered taiga has to a surprising extent been preserved, thanks to the subarctic climate—it is doubtful whether any revelations of great consequence about the impact site will be announced in the future. By now the facts are in; all existing evidence of the June 30, 1908, phenomenon has been photographed, measured, and analyzed—if not explained.

The present situation therefore is that the nature of the explosion is understood but not the cause. Although the case is far from officially closed, the majority of investigators agree that the total weight of the evidence leads inescapably to the conclusion that the holocaust released over Siberia was a high-altitude nuclear explosion of approximately 30 megaton yield. Yet the mystery of the event is not erased or even diminished by such knowledge

—instead, it is deepened. There remains the question, difficult and so far unanswerable by any known natural cause: how could an atomic blast occur in an age when mankind had only a limited grasp of nuclear physics and absolutely no atomic bomb capability?

An essential item of evidence necessary to the final determination of the cause is missing—possibly annihilated by the very atomic conflagration it may have caused. Scientists confronting the Tunguska mystery are faced with a tricky situation similar to that of detectives trying to solve a baffling crime in which the murder weapon is irretrievably lost. The corpse has been carefully autopsied, its organs minutely dissected, the fatal wound probed, and the probable cause of death determined; yet the case cannot be securely concluded nor an arrest made without positive proof establishing the identity of the guilty party.

But if the circumstantial evidence surrounding the case is sufficiently strong and detailed, the missing element of the puzzle may sometimes be deduced or inferred and the facts reconstructed from the shape of the remaining pieces. In attempting to deduce the identity of the enigmatic object of 1908, analysts must cope with unique, unprecedented evidence that is often startling and ambiguous. The most recent hypotheses—such as those regarding a contraterrene particle or a tiny black hole—take us to the outermost edge of our present knowledge of the cosmos into highly speculative, untested areas of astronomy and astrophysics where belief requires a leap of imagination or faith. As the science journalist Walter Sullivan has pointed out, "In contemplating such exotic explanations for the Tunguska event as antirocks or black holes, we realize how far our ideas about nature have gone beyond what we can see, hear and feel."

But all cosmic exploration, actual and theoretical, depends upon such leaps or imaginative journeys beyond the limits of the known. Exploding galaxies, pulsars, and quasi-stellar radio sources (quasars)—to name only a few discoveries of recent astrophysics—are all elements in a gradually unfolding picture of the complex universe that we inhabit. Without fully understanding or being able totally to prove these phenomena, modern astronomers nonetheless are using them to construct a new model or scheme of our universe. "The production of a scheme is a major effort of speculative Reason," wrote the philosopher Alfred North Whitehead in *The Function of Reason*. "It involves imagination far out-running direct observations."

The Siberian riddle, which has perplexed our technology for more than half a century, may yield its meaning when we are able to fit it into an imaginative scheme, an amplified cosmological perspective that, in Whitehead's terms, allows us "to transcend the existing analysis of facts." This cosmology will have to be logical and coherent, responsive to observable facts yet always looking beyond them and finally enabling us to understand them. Kazantsev was the first interpreter of the Siberian disaster to see the problem in these terms; based on his observations at Hiroshima and his ideas about space travel, he formulated an abstract cosmological context to account for evidence that appeared inexplicable. His speculations led to an overthrow of the existing attitude toward the event and to the eventual discovery of fresh facts that had always been there but which had not been seen until his theory made them visible. "One main law which underlies modern progress is that, except for the rarest accidents of chance," Whitehead points out, "thought precedes observation." We see, in other words, only what we are prepared to see."

But even Kazantsev had not been prepared to see far enough. Although he was proved to be correct about the atomic nature of the explosion and his spacecraft theory is now regarded by many as a possibility that cannot be lightly dismissed, his cosmology was definitely too narrow and restricted to explain the possible origin of the Tunguska object. His picture of our solar system has been outmoded by the revelations of our space technology, particularly by NASA and Soviet probes to our closest planets Mars and Venus. Kazantsev envisioned a space ship arriving from Mars, previously considered to be the only planet in the solar system, aside from the earth, that might have environmental conditions favorable to the development of intelligent life. But since then Mariner 9 has given us a closer view of the Red Planet. In 1971, while orbiting Mars, the probe sent back television images of a barren world without water, an environment as inhospitable to higher life forms as the moon, its surface packed with volcanic craters and swept by huge dust storms. In July and September of 1976 two Viking landers, the first crafts specifically designed to search for traces of life on another planet, will set down on Mars and scoop up samples of Martian soil for analysis.

In October 1975 the Soviet interplanetary probes Veneras 9 and 10 landed and transmitted the first photographs of the rock terrain of our other nearest neighbor, Venus, a hot-house globe with a surface temperature of 900 degrees Fahrenheit. Although more space vehicles will be launched, some containing experimental life-detector equipment, the rest of the planets revolving around our sun, in most scientists' opinion, are not likely to have conditions favorable to anything except extremely hardy lower micro-organisms. As intelligent beings, we seem to be alone in our solar system.

In 1973 one of NASA's probes, Pioneer 10, after sending back photographs of Jupiter, headed out of the solar system bearing an aluminum plaque with an engraved message designed for extraterrestrials. The probability of the existence of some form of higher extraterrestrial beings drastically increases if we look past our planetary system which circles an insignificant star on the corner of a vast galaxy containing an estimated one hundred billion stars of all shapes, sizes, and ages. Within our Milky Way galaxy there may be as many as a billion stable suns with their own planetary systems, a million of which may contain warm habitable planets with environments similar to the earth's.

Only fifty years ago we had not yet seen beyond the boundaries of the Milky Way, but we have since then peered into intergalactic space. Surrounding our galaxy, a relatively modest cluster of stars, are billions of other galactic systems, new and old, thriving and expiring, stretching beyond our present powers of detection. The most distant objects detected so far in this great metagalaxy are the quasars, which emit such strong light and radio signals that some astronomers believe they may be newly formed galaxies. Emanating from beyond the quasars are electromagnetic radiations possibly from the prehistory of our metagalaxy. This complex, interrelated metagalactic system, according to some theoretical astronomers and cosmologists, may be only one of billions of other metagalaxies in a universe extending or continuously evolving into infinity.

Can mankind be the sole creatures possessing the faculty of intelligence and the capabilities of a sophisticated technology in all of metagalactic space? Only a few decades ago, when Kazantsev proposed his theory of extraterrestrial visitors, this question was generally regarded as too fanciful for serious scientific investigation. The issue seemed a subject fit only for speculative writers. Today, after American and U.S.S.R. space probes have visited planets in our solar system and radio astronomy has radically reshaped our traditional concepts of the cosmos, there has been a revolutionary change of attitude among many scientists on the possibility of some form of higher extraterrestrial life existing in other solar systems.

The new attitude is reflected in a statement from a report issued in 1972 by the Astronomy Survey Committee of the U. S. National Academy of Sciences:

> Each passing year has seen our estimates of the probability of life in space increase, along with our capabilities for detecting it. More and more scientists feel that contact with other civilizations is no longer something beyond our dreams but a natural event in the history of mankind that will perhaps occur within the lifetime of many of us. The promise is now too great, either to turn away from it or to wait much longer before devoting major resources to a search for other intelligent beings.

. . . In the long run this may be one of science's most important and most profound contributions to mankind and to our civilization.

A landmark study of the subject, *Intelligent Life in the Universe*, by the Soviet astrophysicist I. S. Shklovskii of the Sternberg Astronomical Institute in Moscow and the American exobiologist and astronomer Carl Sagan of Cornell University, authoritatively outlined the current state of scientific knowledge of the biological origins of living organisms on a planet such as the earth and convincingly demonstrated the possibility that the conditions necessary for the evolution of higher life forms may be present on countless planets scattered throughout the cosmos. While stressing the speculative nature of some of their ideas, the authors estimate that in the future not only interstellar communication but flight eventually will be achieved. "Especially allowing for a modicum of scientific and technological progress within the next few centuries," Sagan stated, "I believe that efficient interstellar spaceflight to the farthest reaches of our Galaxy is a feasible objective for humanity. If this is the case, other civilizations, aeons more advanced than ours, must today be plying the spaces between the stars."

That such conclusions about interstellar life and flight are regarded not only as possible but highly probable by an increasingly large portion of the world scientific community was demonstrated by the first international conference, held in 1971 at the Byurakan Astrophysical Observatory in Soviet Armenia, devoted to the question of contact with extraterrestrial life. At this week-long gathering cosponsored by the U. S. National Academy of Sciences and the U.S.S.R. Academy of Sciences, numerous prestigious astronomers, physicists, biologists, anthropologists, linguists, cryptologists, sociologists, and archaeologists from different nations agreed that "the chance of there being extraterrestrial intelligence is much greater than scientists thought possible a few decades ago" and that recent discoveries have transferred some of the problems of detecting other civilizations in the cosmos "from the realm of speculation to a new realm of experiment and observation."

In a 1975 article in *Scientific American,* Carl Sagan and Frank Drake, who headed an earlier detection experiment called Project Ozma in West Virginia, summarized the new awareness of the probability of life existing elsewhere in the universe:

> From the movements of a number of nearby stars we have now detected unseen companion bodies in orbit around them that are about as massive as large planets. From our knowledge of the processes by which life arose here on earth we know that similar processes must be fairly common throughout the universe. Since intelligence and technology have a high survival value it seems likely that primitive life forms on the planets of other stars, evolving over many billions of years, would occasionally develop intelligence, civilization and a high technology. Moreover, we on the earth now possess all the technology necessary for communicating with other civilizations in the depths of space. Indeed, we may now be standing on a threshold about to take the momentous step a planetary society takes but once: first contact with another civilization.

Yet Sagan, Drake, Shklovskii, and most other reputable scientists involved in this experimental search emphasize that what is meant by "extraterrestrial life" is most certainly not the incredibly humanoid "ancient astronauts" conjectured largely from Latin-American artifacts and legends by Erich von Däniken and other writers following his fallacious but popular reasoning in *Chariots of the Gods?* (1970); nor are real extraterrestrials likely to resemble the benevolent blond Venusians or little green Martians claimed to have been sighted by some UFO contactees. These fantasies about alien life reflect an unimaginative, egotistically anthropocentric cosmology in which all living creatures in the universe are merely duplications of us, mirrors of our features and fears.

A more rational, less self-centered cosmological viewpoint, rooted in scientific logic, will admit that extraterrestrials, if they exist, are not likely in any way biologically to resemble what we know on our small planet. "Life, even cellular life, may exist out yonder in the dark," comments naturalist Loren Eisley in his book *The Immense Journey*. "But high or low in nature, it will not wear the shape of man. That shape is the evolutionary product of a strange, long wandering through the attics of the forest roof, and so great are the chances of failure, that nothing precisely and identically human is likely ever to come that way again."

Though life beyond the earth will not assume our shape, we will be related—by the fact that we inhabit the same universe and evolved from the same elements forged in the violent depths of the stars. We are all made, to quote Sagan, "from the dregs of star-stuff." The new cosmological picture emerging today, from ground-based and space astronomy, encourages us to see ourselves and our planet as an infinitesimal but closely related part of an immense, active universe which was created, according to one highly favored cosmic scheme, by an instantaneous "big bang," a cataclysmic nuclear explosion occurring some twelve billion years ago.

In such a universe nothing is fixed or immutable. On a scale of time and space vastly greater than ours, gigantic spiral and elliptical galaxies wheel, burst, and cool in the midst of a cosmos characterized not by stillness and stability but by turbulent events and energetic change. Stars pulsate, flourish, and collapse; bathed in their energy, circling planetary worlds pursue their own peculiar, random destinies, while whatever creatures that happen to be living on their surfaces adopt their own unique evolutionary modes of survival.

Within this considerably expanded cosmological perspective, we can let our imaginations outrun direct observations while staying firmly rooted in current scientific knowledge. We have already accomplished the previously unimaginable—begun the exploration of our solar system.

Our probe, Pioneer 10, is now moving beyond our solar system into interstellar space. Is it fanciful or unreasonable, then, to accept that on some distant planet a civilization may be engaged in similar experimental voyages of discovery, launching their own probes and spacecraft? Once we admit this possibility—accepted by many contemporary scientists—we can then picture the missing element in the puzzling Siberian explosion of 1908.

Loren Eisley offers a provocative image about what may have happened in our past:

> So deep is the conviction that there must be life out there beyond the dark, one thinks that if they are more advanced than ourselves they may come across space at any moment, perhaps in our generation. Later, contemplating the infinity of time, one wonders if perchance their messages came long ago, hurtling into the swamp muck of the steaming coal forests, the bright projectile clambered over by hissing reptiles, and the delicate instruments running mindlessly down with no report.

We can extend our imaginations far out into space and construct an image of the mysterious missile that blew up over the Siberian woods. The image will be necessarily speculative, for no absolute proof exists; yet it will be true to scientific plausibility and to the newly enlarged cosmological view. Let us look at the event with the perspective of almost seventy years of technological progress and research, weighing again the evidence and ordering it in a way that provides a credible—perhaps the most credible—explanation of one of the greatest explosions our world has ever known.

It is the morning of June 30, 1908.

High above the Indian Ocean, a huge object hurtling from space pierces the earth's atmospheric shell. In the almost airless upper altitudes, there is no sound, barely any friction; unimpeded, it races toward the planet.

The object is an extraterrestrial vehicle; its hull is

cylindrical, its mass thousands of tons. Propelled by nuclear fire, the giant craft has come from the depths of interstellar space at a velocity close to the speed of light, then decelerated before orbiting into our planetary system.

Now, rocketing toward the earth, the vehicle is in a state of emergency. Within its propulsion chambers, a malfunction has occurred; the temperature is rapidly rising in the fuel core. Barriers that prevent critical mass, the density necessary for a lethal chain reaction, are overheating and melting.

Eighty miles above the surface, the craft's navigational system steers toward a narrow entry corridor—the same atmospheric passage that many decades later lunar flights from earth will hit for safe re-entry. The entry must be precise, to avoid burning up in the thicker atmosphere or ricocheting back out into the void.

Plunging through the corridor, the craft reduces its velocity. In seconds, as its protective shield strikes the denser air layers, frictional heat rises to 5,000 degrees Fahrenheit, forming a fiery cone of ionized molecules more dazzling than the sun. Within the flaming sheath, the spacecraft glows like a brilliant fireball.

Half of the rotating globe below is in darkness; the sky arching over the other half is cloudless and clear as a glass dome. In a long, sloping trajectory, the craft soars beyond the ocean basin across jagged mountain ridges, steep valleys, vast undulating plains. It navigates directly along a meridian toward the planet's northern horizon, the arctic regions.

A shock wave of highly compressed air is thrust far ahead of the vehicle. The heat shield disintegrates, streaming an incandescent trail of molten particles. Above the northern hemisphere, sensitive optical instruments in the craft register signs of life on the surface.

In central Siberia a deafening roar terrifies the inhabitants of small towns and villages, the only settlements in this remote and deserted area. A powerful ballistic wave pushed before the descending craft strikes the ground.

Trees are leveled, nomad huts blown down, men and animals scattered like specks of dust.

At an altitude of 2 miles, the inhabitants of the luminous spacecraft make a course correction, steering westward over the empty wooded terrain of the Central Siberian Plateau.

The maneuver is their last act.

The barriers separating the fuel cells have melted. The nuclear material reaches a density that is supercritical, and in an instant a chain reaction is triggered.

A fraction of a second later, the spacecraft and its occupants are vaporized in a blinding flash of light.

A towering primordial fire, hotter than the interior of a star, splits the sky in two and sears the landscape below for more than 30 miles.

Then the great fire is gone, leaving behind only a massive column of black clouds that will remain for days in the atmosphere and a scarred, shattered taiga that will forever hide its secret.

Appendixes

A. I. M. Suslov, "In Search of the Great Meteorite of 1908," *Mirovedenie*, Vol. 16, No. 1 (1927), 13–18.

B. L. A. Kulik, "The Problem of the Impact Area of the Tunguska Meteorite of 1908," *Doklady Akad. Nauk SSSR* (A), No. 23 (1927), 399–402.

APPENDIX A

I. M. Suslov, "In Search of the Great Meteorite of 1908," *Mirovedenie*, Vol. 16, No. 1 (1927), 13–18. [Suslov, a Soviet ethnographer, was among the first to interview local inhabitants of the Tunguska region about the disaster.]

In March 1926 I took an expedition to the Chunya River, a right tributary of the Stony Tunguska, at the request of the Committee of Assistance to the Peoples of the North. I also undertook to carry out some scientific investigative work in the region of the Chunya River whose upper course had never been explored.

The expedition took the following route: from the station of Taishet we went by sled to the village of Keshma; from Keshma we reached the Stony Tunguska at the mouth of the Tataria River; then we went further along the Stony Tunguska to the mouth of the Anavar River; the expedition finally reached the Chunya River by traveling on the road to the village of Strelka that was constructed in 1925.

In recent years during my trips in the Angara area I have constantly heard the peasants talk about the fall of the Tunguska meteorite. These stories, although often created from the most highly improbable legends, raise the possibility just the same of finding the fallen meteorite. The most detailed information has come from I. I. Pokrovsky, who carried out statistical work in the Stony Tunguska area during 1908–1909. Unfortunately, the articles about the meteorite by A. V. Voznesensky and S. V. Obruchev, which were published in the first issue of *Mirovedenie* for August 1925, became known to me only a few days before I wrote the present article. As a result, I was unable to make use of the considerable data contained in these reports. This would have been of great help in searching for the meteorite during my trip.

I found similar types of stories in the article by A. V. Voznesensky that I had heard in the area of Keshma. This definitely confirms the accuracy of the peasants' testimony about the ringing of the window glass, loud claps of sound like thunder, a bright light in the sky, and so forth.

At the mouth of the Anavar River I found the tent of the very same Tungus Ilya Potapovich whom S. V. Obruchev had questioned in 1924 on the Tartaria River. With his family lived Akulina, the widow of his brother Ivan. She and her husband had suffered from the fall of the meteorite, and she told me the following about it. Early in the morning, while everyone in the tent was still asleep, the people were suddenly flung into the air. No one was seriously hurt, but Akulina and Ivan lost consciousness. They were utterly terrified and for a long time did not know what had happened. Then they saw the forest blazing around them with many fallen trees. There was also a great noise.

At the village of Strelka I met the old Tungus Vasily Okhchen who at the time of the fall of the meteorite lived in the tent of Ivan and Akulina. I questioned him two weeks after I had talked with her, and he told me the same story. The only difference was that Vasily had been sleeping at the moment when the tent was torn away and had been thrown to the side by a powerful jolt. He had not lost consciousness. He said that he heard an unbelievably loud and continuous thunder; the ground shook, burning trees fell, and all around there was smoke and haze. Soon the thunder stopped, the wind ceased, but the forest continued to burn. All three of the Tungus went out to search for the reindeer which had run away during the catastrophe. But they were unable to find many of the reindeer from the herd.

At the time of the fall of the meteorite Akulina's hut was located at the mouth of the Diliushmo River where it flows into the Khushmo River. At the same time Ilya Potapovich lived on the Tartaria River. When the meteorite fell, he heard only long peals of thunder and felt the earth shaking.

Afterward, while they were out hunting for squirrels, Ilya Potapovich and Akulina found on the northeastern slope on the Lakura Ridge, close to the source of the Makirta River, a dry stream—a deep fissure which ends in a large pit filled up with earth. According to them, at the present time both the fissure and the pit have been grown over with young trees.

On the middle course of the Avarkita River at the time of the catastrophe stood the tent of the children of the deceased Tungus Podyga: Chekaren, Chuchancha, and Malega. I met them at the village of Strelka. They said that they had been startled by loud rumbling noises. Everywhere they heard great crashing sounds and felt the earth shake. A terrible storm, so great that it was difficult to stand upright in it, blew down the trees near their hut. In the distant north, a large cloud formed which they thought was smoke.

The Tungus Andrei Onkoul, who now lives at the crest of the Taimura River, said that north of the Lakura Ridge, approximately halfway between the Kimchu and Khushmo Rivers, he saw a huge pit which the Tungus had earlier known nothing about. This pit was also grown over with young trees.

An entire group of Tungus said that a fire suddenly flew into the bank of the Chambé River, slightly below the mouth of the Khushmo River, and quickly burned up two hundred reindeer that belonged to the Tungus Stephan Ilich Onkoul from the tribe of Kurkagyr. His storehouses, which were filled with sacks of flour and with domestic goods, were knocked down and completely destroyed.

Tungus stories about any event must always be carefully checked, even more so a supernatural one, and as in this particular case when many of them lost consciousness and were hurt by the powerful force of the wind. Therefore I took advantage of a meeting with sixty Tungus that occurred from June 1 to June 4, 1926, at the village of Strelka, to carry out an official interview of all those present. This interview also provides extremely valuable information.

Such expressions were heard as "the forest was crushed," "the storehouses were destroyed," "the reindeer were annihilated," "people were injured," "the dogs were killed," "the taiga was flattened," "trees fell from the summit of Nerbogachen (a mountain to the northeast).". . . "this brought with it a disease for the reindeer, specifically scabs, that had never appeared before the fire came." The stories of Ilya Potapovich, Akulina, Vasily Okhchen, Andrei Onkoul, Chuchancha, and Chekaren were confirmed by everyone.

It is interesting to note that they all willingly answered my questions about the details of what they had suffered and at the same time were ready to show any place connected with the fall of the meteorite. They readily agreed to my request to draw a map of the area of the catastrophe. Ilya Potapovich drew the map with colored pencils, and a group of Tungus made corrections. . . .

Translated by John W. Atwell

APPENDIX B

L. A. Kulik, "The Problem of the Impact Area of the Tunguska Meteorite of 1908," *Doklady Akád. Nauk SSSR* (A), No. 23 (1927), 399–402.
[This scientific report, written shortly after Kulik returned from his first expedition to the Tunguska region, contains his initial impressions of the blast site and three important eyewitness accounts.]

In February 1927 the Academy of Sciences of the U.S.S.R. equipped an expedition, headed by the author, for the investigation of the impact area of the meteorite of June 30, 1908. The area of the fall was the basin of the upper course of the Stony Tunguska River.

At the end of March 1927, I began to reconnoiter the area north of the Stony Tunguska River. My base was the Vanavara trading station which lies on this river close to longitude 72 degrees east. After a number of attempts to penetrate the taiga north of this river by probing in a northwesterly direction by raft, I finally reached the central area of the fall in June. I made a summary investigation of this area and its surroundings.

In view of the absence here of any astronomical points for hundreds of kilometers around and the complete lack of any maps for the region, I can only estimate the location as approximately latitude 61 degrees north and longitude 71 degrees east—about 100 kilometers northwest of the Vanavara trading station.

The central area of the fall is several kilometers in diameter and lies on the watershed between the basin of the Chunya River and the plateau of the Stony Tunguska. The area is a huge hollow surrounded by an amphitheater of ridges and mountain peaks. These mountains are touched on the south by the Khushmo River, the right tributary of the Chambé River, which flows from northwest to southeast. The Chambé flows into the Stony Tunguska about 30 kilometers below the Vanavara trading station. This system of tributaries was my main route from Vanavara in both directions. The amphitheater contains hills, ridges, various peaks, tundra, swamps, lakes, and streams. Not long ago, according to the local inhabitants, this area was a typical part of the taiga. Now, both within and outside of the hollow, everything has been practically destroyed. The trees lie in rows on the ground, without branches or bark, in the direction opposite to the center of the fall. This peculiar "fan" pattern of fallen trees can be seen very well from some of the heights that form the peripheral ring of trees. Nevertheless, here and there the trees remain standing (usually without branches or bark). In addition, in certain places there are still some small patches of green trees. These are rare, however, and can be easily explained. All of the previous growth, both within the hollow and in the surrounding mountains and also in the zone that extends for some kilometers around, show the traces of an intense fire. This is unlike the usual fire and is found both on the fallen as well as on the standing trees. It is also evident on the remains of bushes and the summits and slopes of the mountains, on the tundra and on the isolated islands in the water covered swamp. The area that shows the traces of the fire is some tens of kilometers in diameter. The central part of this burned area is several kilometers in diameter. In the part of the area that is covered with plant growth and with the trees of the taiga there is evidence of some kind of pressure that came from the side and produced creased depressions in the ground several meters deep. In general, these stretch perpendicularly in a northeasterly direction. In addition, the area is dotted with dozens of freshly formed craters of various diameters, from a few meters to tens of meters with depths of tens of meters also. The sides of these craters are usually steep although sloping ones are also found. The bottoms of the craters are flat and marshy with an occasional elevation in the center. At the northeastern end of one of the sections of the tundra the swamp has replaced the tundra at the foot of the mountains for several tens of meters. On the other side, in the southwestern corner of the amphitheater, the swamp ends in a chaotic conglomeration of wet cover.

A detailed excavation of one small crater (1.5 to 2 meters in diameter) filled with swamp water, showed that (1) water can be removed from this type of crater by pails; (2) that frozen peat can be found on the bottom of the crater at the end of June; (3) that the layer of water in a crater measures 30 centimeters and the layer of liquid silt beneath it also measures 30 centimeters; (4) that the rise of the water in a twelve-hour period did not exceed 30 centimeters, i.e., half of the original quantity.

To what I have already described I must add the interesting evidence about the fall that I gathered from eyewitnesses. The peasant S. B. Semenov gave me the following in written form: "This took place in June 1908 at eight o'clock in the morning. At this time I was living at the Vanavara trading station on the Stony Tunguska River. I was working on my hut. I was sitting on my porch facing north when suddenly, to the northwest, there appeared a great flash of light. There was so much heat that I was no longer able to remain where I was—my shirt almost burned off my back. I saw a huge fireball that covered an enormous part of the sky. I only had a moment to note the size of it. Afterward it became dark and at the same time I felt an explosion that threw me several feet from the porch. I lost consciousness for a few moments and when I came to I heard a noise that shook the whole house and nearly moved it off its foundation. The glass and the framing of the house shattered and in the middle of the area where the hut stands a strip of ground split apart. The barn door also broke, although the lock was unharmed."

Another peasant, P. P. Kosolapov, told me in a personal interview on March 30, 1927, that in June 1908 at eight o'clock in the morning he was at the same trading station. He needed some nails; not finding any, he went out into the yard. He began to pull nails out of the frame of the window with a pair of tongs. Suddenly a powerful heat began to burn his ears; he put his hands over them and thought that the roof was on fire. He raised his head and asked Semenov if he had seen anything. "How could one help but see it?" Semenov answered. "I felt as though I had been seized by the heat." Kosolapov had just gone into the house and had started to sit down to work when there was a great clap of thunder and the sod poured from the ceiling, a door flew off the Russian stove, and a piece of window glass fell into the room. When it was quieter Kosolapov went into the yard, but he did not see anything else.

Finally the Tungus Ilya Potapovich [Liuchetkan] told me in an interview on April 16, 1927, that the center of the firestorm was the pasture land of the Tungus Vasily Ilich, the uncle of his first wife. Vasily Ilich was a wealthy Tungus who owned about fifteen hundred reindeer. In the same area he owned a number of storehouses where he kept clothes, household goods, reindeer harnesses, and so forth. Most of the reindeer were in the mountains in the region of the Khushmo River. The fire came by and destroyed the forest, the reindeer, and the storehouses. Afterward, when the Tungus went in search of the herd, they found only charred reindeer carcasses. Nothing remained of the storehouses; everything had burned up and melted—clothes, utensils, reindeer harnesses, dishes, and samovars. Only a few pails remained. All of these places are known to the brothers of Vasily Ilich—Burucha and Mugocha.

Other Tungus and Russian witnesses confirmed that (1) the trees in the center of the fall area lie in a fan pattern and the tops point away from the center of the fall; (2) before the fall this was a normal green taiga; (3) there have been no other fires in this area either before or after the fall and the fire that it caused.

The investigation that I conducted was brief because of a lack of time and supplies and also because of a lack of technical equipment.

A further detailed investigation of the area of the fall is very necessary.

In literature on meteorites, if the evidence from chronicles is discounted, there have never been any descriptions of a firestorm produced by the fall of a meteorite although there have been indications of firestorms in the past. In this particular fall we have the first example of a great firestorm which can and unquestionably must be studied in detail. Moreover, this is one of those apparently rare occasions when a large meteorite reaches the earth's surface accompanied by the cloud of burning gases that surround it until it reaches our atmosphere. Finally, the character of the dispersal of the craters in the area of the fall demands more detailed study since this is an unusual fall both in terms of its mass and its speed. In addition, this study will help in choosing a more advantageous spot for future excavations. The best method would be the taking of photographs by a seaplane. It would also be desirable to make a magnetic or electrical check of the crater area. The height of the plateau and the surrounding mountains should be determined as well as the fixing of several astronomical points in order to check the photographs taken in the area.

Translated by John W. Atwell

Bibliography

Aside from the New York *Times* and the London *Times*, whose pages we frequently consulted, the following are our basic reference sources in Russian and English.

RUSSIAN PUBLICATIONS

(Doklady Akad. Nauk SSSR [*Papers of the Academy of Sciences, U.S.S.R.*]; *Mirovedenie* [*World Knowledge*]; *Priroda* [*Nature*]; *Znaniye-Sila* [*Knowledge Is Strength*].

Astapovich, I. S. "New Investigations of the Fall of the Great Siberian Meteorite of 30 June 1908," *Priroda,* No. 9 (1935), 70–72.

——. "The Great Tunguska Meteorite," *Priroda,* No. 2 (1951), 23–32.

Bronstein, V. A. "The Problem of the Movement of the Tunguska Meteorite in the Atmosphere," *Meteoritika,* Issue 20 (1961), 72–86.

Fesenkov, V. G. "Atmospheric Turbidity Caused by the Fall of the Tunguska Meteorite," *Meteoritika,* Issue 6 (1949), 8–12.

Florensky, K. P. "Preliminary Results from the 1961 Combined Tunguska Meteorite Expedition," *Meteoritika*, Issue 23 (1963), 3–29.

Kazantsev, A. P. A. *Guest from the Cosmos: Tales and Stories*. Moscow, 1963.

Kulik, L. A. "The Lost Filimonovo Meteorite of 1908," *Mirovedenie*, Vol. 10, No. 1 (1921), 74–75.

——. "The Search for the Lost Filimonovo Meteorite of 1908," *Mirovedenie*, Vol. 11, No. 1 (1922), 80.

——. "History of the Bolide of June 30, 1908,' *Doklady Akad. Nauk SSSR* (A), No. 23 (1927), 393–98. [English translation in *Popular Astronomy*, Vol. 43, No. 8 (1935), 499–504.]

——. "The Problem of the Impact Area of the Tunguska Meteorite of 1908," *Doklady Akad. Nauk SSSR* (A), No. 23 (1927), 399–402.

——. "Preliminary Results of Meteorite Expeditions in the Decade 1921–31," *Papers of the Lomonosov Institute. Akad. Nauk SSSR*, No. 2 (1933), 73–81. [English translation in *Popular Astronomy*, Vol. 44, No. 4 (1936), 215–20.]

——. "Data on the Tunguska Meteorite Obtained by 1939," *Doklady Akad. Nauk SSSR*, No. 8 (1939), 520–24. [English translation (excerpt), with commentary by E. L. Krinov, in *Source Book in Astronomy 1900–1950*, Harlow Shapley (ed.). Cambridge: Harvard University Press, 1960.]

Levin, B. "The Problem of the Velocity and Orbit of the Tunguska Meteorite," *Meteoritika*, Issue 11 (1954), 132–36.

Obruchev, S. V. "The Place Where the Great Meteorite of 1908 Fell," *Mirovedenie*, Vol. 14, No. 1 (1925), 38–40.

——. "More About the Place Where the Tungus Meteorite Fell," *Priroda*, No. 12 (1951), 36–38.

Suslov, I. M. "In Search of the Great Meteorite of 1908," *Mirovedenie*, Vol. 16, No. 1 (1927), 13–18.

Tronov, M. V. (ed.). *The Problem of the Tungus Meteorite*. Tomsk: Tomsk University Press, 1967.

Voznesensky, A. V. "The Fall of the Meteorite of June 30, 1908," *Mirovedenie*, Vol. 14, No. 1 (1925), 25–38.

Zigel, Felix. "Nuclear Explosion over the Taiga: Study of the Tunguska Meteorite," *Znaniye-Sila*, No. 12 (1961), 24–

27. [English translation in Joint Publications Research Service, Washington, D.C., JPRS-13480 (April, 1962).]

——. "Unidentified Flying Objects," *Soviet Life* (February 1968, 27–29.

Zolotov, A. V. "Problems of the Dust Structure of the Tungus Cosmic Body," in M. V. Tronov (ed.). *The Problem of the Tungus Meteorite*. Tomsk: Tomsk University Press, 1967, pp. 173–87.

——. "The Relationship of the Geomagnetic Effect Caused by an Atomic Explosion in the Air with the Blast," in M. V. Tronov (ed.). *The Problem of the Tungus Meteorite*. Tomsk: Tomsk University Press, 1967, pp. 162–68.

Zotkin, I. T., and M. A. Tsikulin. "The Scale Reproduction of the Explosion of the Tungus Meteorite," *Doklady Akad. Nauk SSSR*, Vol. 167, No. 1 (1966), 59–62.

ENGLISH PUBLICATIONS

Alfvén, Hannes. *Worlds-Antiworlds*. San Francisco and London: W. H. Freeman, 1966.

Astronomy and Astrophysics for the 1970's. Report of the Astronomy Survey Committee, Vol. 1. Washington, D.C.: National Academy of Sciences, 1972.

Berendzen, Richard (ed.). *Life Beyond Earth & the Mind of Man*. Symposium held at Boston University on November 20, 1972. Washington, D.C.: NASA, 1973.

Brown, Harrison (ed.). *A Bibliography on Meteorites*. Chicago: University of Chicago Press, 1953.

Brown, Peter Lancaster. *Comets, Meteorites and Men*. New York: Taplinger, 1974.

Chapman, Sidney, and Attia A. Ashour. "Meteor Geomagnetic Effects," *Smithsonian Contributions to Astrophysics*, Vol. 8, No. 7 (1965), 181-97.

Collins, Michael. *Carrying the Fire: An Astronaut's Journeys*. New York: Farrar, Straus & Giroux, 1974.

Condon, Edward U. (project director). *Scientific Study of Unidentified Flying Objects*. New York: E. P. Dutton, 1969.

Cowan, C., C. R. Atluri, and W. F. Libby. "Possible Anti-Matter Content of the Tunguska Meteor of 1908," *Nature* (London), Vol. 206, No. 4987 (1965), 861–65.

Crowther, J. G. "More About the Great Siberian Meteorite,"

Scientific American, Vol. 144, No. 5 (1931), 314–17.

Digby, [George] Bassett. *Mammoths and Mammoth Hunting.* New York: D. Appleton, 1926.

——. *Tigers, Gold and Witch Doctors.* New York: Harcourt, Brace, 1928.

Eisley, Loren. *The Immense Journey.* New York: Random House, 1946.

Florensky, K. P. "Did a Comet Collide with the Earth in 1908?" *Sky and Telescope*, Vol. 25, No. 26 (1963), 268–69.

Groves, Leslie R. *Now It Can Be Told: The Story of the Manhattan Project.* New York: Harper & Row, 1962.

Hellman, Lillian (ed.). *The Selected Letters of Anton Chekhov.* New York: Farrar, Straus & Cudahy, 1955.

Heninger, S. K. *A Handbook of Renaissance Meteorology.* Durham, N.C.; Duke University Press, 1960.

Hersey, John. *Hiroshima.* New York: Knopf, 1946.

Hobana, Ion, and Julien Weverbergh. *UFO's from Behind the Iron Curtain.* New York: Bantam, 1975.

Hoyle, Fred. *Man in the Universe.* New York: Columbia University Press, 1966.

Jackson, A. A., and M. P. Ryan. "Was the Tungus Event Due to a Black Hole?" *Nature* (London), Vol. 245, No. 5420 (1973), 88–89.

Krinov, E. L. *Principles of Meteorites.* New York: Pergamon, 1960.

——. *Giant Meteorites.* New York: Pergamon, 1966. [Chapter entitled "The Tunguska Meteorite," pp. 125–265.]

Lamont, Lansing. *Day of Trinity.* New York: Atheneum, 1965.

Lang, Daniel. *From Hiroshima to the Moon.* New York: Simon and Schuster, 1959.

La Paz, Lincoln. "Meteorite Craters and the Hypothesis of the Existence of Contraterrene Meteorites." *Contributions of the Society for Research on Meteorites*, Vol. 2, No. 4 (1941), 244–47.

Laurence, William L. *Dawn Over Zero: The Story of the Atomic Bomb.* New York: Knopf, 1946.

Lawrence, David. "What Hath Man Wrought!" *U. S. News & World Report* (August 17, 1945).

Lengyel, Emil. *Siberia.* New York: Random House, 1943.

Lewis, Richard S. *Appointment on the Moon: The Inside*

Story of America's Space Venture. New York: Viking, 1968.

Ley, Willy. *Watchers of the Skies.* London: Sidgwick & Jackson, 1964.

Liebow, Averill A. *Encounter with Disaster: A Medical Diary of Hiroshima, 1945.* New York: Norton, 1971.

Manchester, William. *The Glory and the Dream: A Narrative History of America, 1932–1972.* Boston: Little, Brown, 1974.

Mowatt, Farley. *The Siberians.* Boston: Little, Brown, 1970.

Maxwell, Robert (ed.). *Information U.S.S.R.* New York: Pergamon, 1962.

Oliver, Charles P. "The Great Siberian Meteorite," *Scientific American,* Vol. 139, No. 1 (1928), 42–44.

Osman, Tony. "Into the Black Hole," *Sunday Times Magazine* (London), (August 19, 1973), 8–12.

Parry, Albert. *Russia's Rockets and Missiles.* London: Macmillan, 1961. [Chapter entitled "The Tungus Mystery: Was It a Spaceship?" pp. 248–67.]

Sagan, Carl. *The Cosmic Connection: An Extraterrestrial Perspective.* Garden City, N.Y.: Doubleday, 1973.

——, and Frank Drake. "The Search for Extraterrestrial Intelligence," *Scientific Amercian,* Vol. 232, No. 5 (1975), 80–89.

Saslaw, William C., and Kenneth C. Jacobs (eds.). The *Emerging Universe: Essays on Contemporary Astronomy.* Charlottesville: University of Virginia Press, 1972.

Sears, Thad P. *The Physician in Atomic Defense.* Chicago: Year Book Publishers, 1953.

Semyonov, Yuri. *Siberia: Its Conquest and Development.* Baltimore: Helicon Press, 1963.

Shklovskii, I. S., and Carl Sagan. *Intelligent Life in the Universe.* New York: Holden-Day, 1966.

Sullivan, Walter. "Curiouser and Curiouser: A Hole in the Sky," New York *Times Magazine* (July 14, 1974), 10, 24–35.

Taylor, John. G. *Black Holes: The End of the Universe?* New York: Random House, 1973.

Watson, Fletcher G. *Between the Planets.* Philadelphia: Blakiston, 1941.

"What a Meteor Did to Siberia," *The Literary Digest* (March 16, 1929), 33–34.

Whipple, F. J. W. "The Great Siberian Meteor and the Waves It Produced," *Quarterly Journal of the Royal Meteorological Society* (London), Vol. 16, No. 236 (1930), 287–304.

Whitehead, Alfred North. *The Function of Reason.* Boston: Beacon Press, 1958.

About The Authors

Thomas Atkins was educated at Duke and Yale Universities. Currently Associate Professor and Chairman of the Theatre Arts Department at Hollins College in Virginia, he edits an international magazine and is the author of five books.

John Baxter was born and raised in Sydney, Australia. A regular contributor to the London *Times* and the *Sunday Times Magazine*, his most recent book is *Stunt: The Story of the Great Movie Stuntmen.*